Voltammetry

Edited by Nobanathi Wendy Maxakato,
Sandile Surprise Gwebu
and Gugu Hlengiwe Mhlongo

Published in London, United Kingdom

IntechOpen

Supporting open minds since 2005

Voltammetry
http://dx.doi.org/10.5772/intechopen.75223
Edited by Nobanathi Wendy Maxakato, Sandile Surprise Gwebu and Gugu Hlengiwe Mhlongo

Contributors
Koichi Jeremiah Aoki, Jingyuan Chen, Santhy Wyantuti, Safri Ishmayana, Diana Rakhmawaty Eddy, Yeni W Hartati, Keiichi Sakamoto, Rajendrachari Shashanka, Kenneth Brown, Pipat Chooto

First published in London, United Kingdom, 2019 by IntechOpen
IntechOpen is the global imprint of INTECHOPEN LIMITED, registered in England and Wales, registration number: 11086078, The Shard, 25th floor, 32 London Bridge Street
London, SE19SG – United Kingdom
Printed in Croatia

British Library Cataloguing-in-Publication Data
A catalogue record for this book is available from the British Library

Additional hard and PDF copies can be obtained from orders@intechopen.com

Voltammetry
Edited by Nobanathi Wendy Maxakato, Sandile Surprise Gwebu and Gugu Hlengiwe Mhlongo
p. cm.
Print ISBN 978-1-78985-323-0
Online ISBN 978-1-78985-324-7
eBook (PDF) ISBN 978-1-83880-847-1

For EU product safety concerns:
IN TECH d.o.o., Prolaz Marije Krucifikse Kozulić 3, 51000 Rijeka, Croatia, info@intechopen.com or visit our website at intechopen.com.

Meet the editors

Dr. Nobanathi Wendy Maxakato is a Senior Lecturer and Researcher at the University of Johannesburg in the Department of Applied Chemistry. Her research field is Electrochemistry with a special interest in Fuel Cells. Her other fields of expertise are Nanotechnology and Materials Science.

Sandile Surprise Gwebu is a Researcher at the University of Johannesburg in the Department of Applied Chemistry. Her research work includes the preparation and electrochemical characterization of alcohol fuel cell electrocatalysts.

Prof. Gugu Hlengiwe Mhlongo is a Physicist. She is a Senior Researcher and characterization facility Manager at CSIR- National Centre for Nano-Structured Materials (NCNSM), South Africa. Her research interests include synthesis and characterization of nanostructured composites for use in numerous fields including energy, health, and safety amongst many.

Contents

Preface

This Edited Volume is a collection of reviewed and relevant research chapters, concerning the developments within the "Voltammetry" field of study. The book includes scholarly contributions by various authors and edited by a group of experts pertinent to Chemistry. Each contribution comes as a separate chapter complete in itself but directly related to the book's topics and objectives.

The book is divided in 6 chapters: "Cyclic Voltammetry and Its Applications", "Tips of Voltammetry", "Cyclic Voltammetry of Phthalocyanines", "Electrochemical Preparation and Characterization of Chemically Modified Electrodes", "Investigation of Electrochemical Pitting Corrosion by Linear Sweep Voltammetry: A Fast and Robust Approach" and "Carbon-Based Electrode Application for Determination and Differentiation of Chromium Ion Species Using Voltammetric Method."

The target audience comprises of all the players in academic communities: scientists, university and college professors, research professionals, students and users of academic libraries.

IntechOpen

Chapter 1

Cyclic Voltammetry and Its Applications

Pipat Chooto

Abstract

Cyclic voltammetry is a versatile method for scientific investigation and innovation due to the fact that most processes involve electron transfer, which makes them be able to be monitored by this technique. Its uses cover characterization, synthesis, mechanisms, and analysis. In all applications, the technique can work well with a large variety of compounds including organic, inorganic, polymer, films, and semiconductors, among others. Furthermore, the method operates satisfactorily whether in a direct or an indirect approach. It can be considered to be an essential part at the very beginning that leads further to the grand project. As an analytical tool, it plays an important role in not only chemistry but also other involving areas. This review sheds the light on the way the technique proves to become one of the important instruments in research, development, and application labs.

Keywords: cyclic voltammetry, electrochemistry, electrochemical applications, characterizations, analysis

1. Introduction

Cyclic voltammetry has long been well known for its versatile applications in a number of areas involving electron transfer process, both directly and indirectly. It should be recognized at first that the techniques are normally not the main or the most important one in the research. Rather, their data are really useful in fulfilling the work as an essential complementary technique. It has been well applied in monitoring electrochemical behavior of a great variety of substances due to the fact that it can provide the insights into the relationship of structure, potential, and characteristic activities. The technique is prominent with its advantages of simplicity, sensitivity, speed, and low costs, among others, which results in a wide range of applications so far. Only important cases of previous studies are presented. Brief information for examples of recent applications is focused here along with corresponding references for further studies in specific area of interest. Furthermore, due to the fact that there have been a large variety of applications, it is not easy to categorize. Therefore, the content has been arranged to suit its further use, vision extension, and initiation of new research. Finally, it has to be emphasized that at this point, readers are supposed to be familiar with basics of this technique as well as involving theories (for example, ref. [29] and [30]). If not, available pertaining literature can be very useful. In discussing the application of cyclic voltammetry, we are going to talk about the three main uses of cyclic voltammetry first for different types of compounds, then continue with involving areas that cyclic voltammetry is considered useful. CV is used starting from here instead of cyclic voltammetry for convenience.

2. Classifications of usage

2.1 Characterization and synthesis

General concepts on using CV for the characterization of polymers as well as inorganic materials can be found elsewhere [1].

2.1.1 Organic compounds and polymers

CV has been proved to be helpful in characterizing xanthone, a bioactive compound, revealing that catechol is the key moiety to make it an effective scavenger for reactive species [2]. CV in methanol of antibiotic named amoxicillin revealed quasi reversible behavior and the change of redox potential upon the interactions with both metal ions and amino acids, which helps guiding in prescribing the medicine that they should not to be used together with the antibiotic [3–5]. Electrochemical behavior of ibuprofen and its degradation could be monitored closely by CV and found to be different for various types of electrode; for example, adsorption was observed on silver composite electrode, providing the insights into selecting the suitable electrode for removing this pollutant [6]. Together with EIS, CV is also found to be very useful in characterizing redox designing biosensors their behaviors depend on diffusion and capacitance, which might well shed the light on designing biosensors [7]. N-dodecylacrylamide (DDA) can form copolymer well with ferrocene derivatives on ITO to be applicable in the catalytic and sensor devices [8]. Its electron transfer characteristic is determined by the formation of monolayer or multilayer on the surface. A polymer-modified electrode namely poly(vinylferrocenium) on Pt was investigated and developed successfully by CV technique to make it applicable in analyzing organic compounds such as adenine [9]. The film of ruthenium complex can be synthesized via electropolymerization and then modified onto Pt electrode as a sensor for hydrazine with the characterization by CV [10]. Due to the fact that pyrrole is electroactive, the behavior of various types of polypyrrole can be conveniently followed by electrochemical methods, especially CV, and they were widely investigated in the 1990s and beyond [11–14]. CV could also be applied well in characterizing polyaniline to get suggestion in film preparations that its kinetic behaviors have the influence from the dopant acid [15]. Our work has also proved that CV can be used in electrografting born-doped diamond (BDD) surface via CV cycling without the use of reducing agent to prepare ex-situ working electrode for Cd determination [16]. With its advantage of speed, fast scan CV has been used to recommend the use of paraffin wax as a sealant instead of epoxy in making electrode for both in vivo and in vitro applications [17]. For natural polymers, CV has paved its way in analyzing specific electroactive compounds such as quinones in natural rubbers, which is very helpful in understanding their properties [18]. It is therefore clear that CV will still be greatly used for organic synthesis and characterizations for a number of years to come.

2.1.2 Metals, metal complexes, and inorganic compounds

Due to the fact that complex formation involves electron transfer, understanding the behavior of metal-ligand interaction can be studied well by CV and proven satisfactorily for simulations for both higher and lower ligand concentrations as well as for obtaining overall formation constants [19]. Our contribution has shown the use of CV to support the results from X-ray crystallography to explain the behavior of copper-thiourea-halide complexes [20]. Redox potential and reversibility obtained by CV have been proved to be useful in studying metalloenzymes with cofactor of molybdenum complexes as a model to describe its physiological role [21]. The characterizations of metal complexes have also been indicated to help investigate the catalytic activities of complexes for the removal of toxic phenolic substances, especially dyes [22]. At quantum level, redox potentials from CV can be used in calculating LUMO and HOMO as well as the gaps of new organic semiconductors, which are applicable in everyday life [23]. CV has been proved to be advantageous in characterizing quantum dots (QDs), particularly in terms of their stabilities in vivo [24]. CV has been shown to be effective in following molecular imprinting of copper complexes [25]. CV can also be used in tracing activated iridium oxide film (AIROF) microelectrode via the variation of scan rate during the steps of fabrications and animal implantation [26]. In particular, CV from gold rotating disk electrode (Au RDE) has also been proved to be useful in determining surface coverage of silica microparticle monolayer and revealing that the adsorption is stronger with increasing diameter and surface coverage, which sheds the light on the deposition of a variety of particles onto the surface [27]. In terms of fuel cells, CV has been used to characterize the synthesized Pt-Ru catalysts and their catalytic activities [28]. CV has been used in the characterization of metal hydride. Using NaOH instead of KOH reduces hydrogen adsorption/desorption, resulting in better performance of nickel-metal hydride batteries [29].

2.2 Mechanism

Generally, CV has been widely recognized to be able to identify the mechanism of coupled chemical reaction [30–32]. For recent applications, cyclic voltammetry was applied together with voltabsorptometry in understanding the reaction of lumazine [33]. Special technique of fast Fourier transform continuous cyclic voltammetry was developed to be used well in investigating the mechanism of 4-nitrocatechol and its electrochemical synthesis [34]. This technique is also used in investigating proteolysis of milk protein by the research group in the same country [35]. The nontriangular waveform of CV has also been adopted with the advantage of convenient determination of transfer coefficient and electron transfer rate constant [36]. Additionally, staircase was compared with linear scan to assist in investigating adsorbed species [37]. Surprisingly, CV can also be used in studying the mechanism of surface coating [38]. Cyclic voltammetry was used together with potentiometry in understanding interactions between nicotine and metal ions [39]. The mechanism of electropolymerization and catalytic reaction of phenol red can be explained well by CV in order to analyze acetaminophen and dopamine [40]. The technique is also helpful in interpreting Ni electrodeposition for extending methanol oxidation [41]. Additionally, it can be used in understanding the mechanism of synthesizing graphene-coated graphene-coated electrode conveniently and fast in only one step [42]. The technique has been proved to be a helpful tool for mechanistic modeling for biosensors [43].

2.3 Analysis

CV can be used extensively for a wide range of both inorganic and organic compounds. In addition, it can be applied indirectly to analyze other types of characteristics, for example, microorganisms or antioxidant properties. Here the focus is on recent examples for analysis applications.

2.3.1 Organic compounds

For the analysis of organic compounds, cyclic voltammetry can be considered versatile in terms of a great variety of compounds. The technique has been used in the detection of glucose with graphene modified electrode [44]. Due to the overlapping of redox peaks in cyclic voltammetry, mathematical operation such as deconvolution has been applied in analyzing electroactive species including explosives, especially nitro compounds. With modern technology in data analysis, the techniques can be used in the determination of pesticides [45]. Separation technology, for example, electromembrane is also helpful in eliminating interferences in diclofenac analysis [46]. Carbon black-modified GCE was used in estradiol by voltammetry in cooperation with FIA and amperometry [47]. Another important natural compound that has been widely used is curcumin, and its characteristics from CV can shed the light for its analysis in various corresponding products [48, 49]. CV was also proved to be supersensitive for hydrazine sensing [50]. Boron-doped diamond also exhibits its performance as an electrode with superb sensitivity for metronidazole in the analysis by CV [51], whereas bismuth electrode proved its specific characteristic in monitoring thiol content in fossil fuel [52]. The determination of 2,4,6-trichloroanisole in cork stoppers to reduce the wine defect can be accomplished satisfactorily with CV [53]. Recently, CV has played very important role in the analysis of juice [54] and wine [55] with particular analytes such as phenolic compounds. The analysis by CV helps in investigating perspective ionic liquids (ILs) to be used as solvents and electrolytes with their specific properties of cation and anion dependence [56]. It is also important to state here that CV can be used in analysis with the increase or decrease of the current signals of indicator compounds such as quinones [57] or ferricyanide [58]. Analyses of organic compounds by CV are sure to appear extensively in the near future in the analytical literature.

2.3.2 Inorganic compounds

Cyclic voltammetry can be used well with numerous inorganic compounds. It is normally applied to simple inorganic compounds probably because their electrochemical behavior is less complicated. It is worth mentioning here that metal ions are normally monitored by stripping voltammetry, so it is not going to delve into details here. Certainly, voltammetry is sensitive to metal ions, but its specificity can be enhanced by electrode modifications and stripping potential. Those who are interested in this area are suggested to get more details elsewhere [59–61]. An example of application of cyclic voltammetry to complicated inorganic compounds is the case of nanostructured iron oxides [62]. Additionally, corrosive impurities can be monitored closely in molten chloride by CV, which is very useful for energy storage [63]. With the use of chitosan and silver nanocomposite modified electrode, hydrogen peroxide can be detected [64], and this can be further extended to the determination of other compounds involving hydrogen peroxide such as glucose.

2.3.3 Other aspects of analysis

With electrochemical characteristics of CV, various analytical aspects of applications in analysis can be achieved as well. It was used in the valuation of antioxidant potential, which is very important in making wine [65]. The technique, along with EIS, CV can be used in assessing emulsion stability [66]. Furthermore, CV can be used in determining HOMO and LUMO levels with an acceptable accuracy [67]. Surprisingly, it can be applied in monitoring the formation of biofilms [68]. With its sensitivity for antioxidant, CV can be used in classifying black tea [69]. It is clear that indirect uses of CV can be used in the future in a variety of applications.

3. Areas of applications

3.1 Chemistry

Electron transfer definitely plays an important role in all reactions; therefore, electrochemistry and CV in particular have an essential part in all aspects of

Areas of chemistry	Typical applications
Analytical chemistry	Preparation and synthesis
	Analysis
	Mechanism
Inorganic chemistry	Metal interactions and reactions
	Structure characterizations
	Analysis
Organic chemistry	Synthesis
	Analysis
	Characterization
	Mechanism
Physical chemistry	Thermodynamic studies
	Kinetic studies
	Theoretical equations
	Surface and adsorption
	Analysis
	Mechanism
Biochemistry (and biology)	Metabolic processes
	Continuous monitoring
	In vivo and in vitro analysis
	Kinetic of enzymes
	Biosensors

Table 1.
Summary of the main uses of CV in chemistry.

chemistry as summarized in **Table 1**. Typical examples have already been mentioned in the previous text.

3.2 Other areas

Due to the fact that CV provides essential information for better understanding and the instrument is quite cost-effective now, thanks to modern technology, not to mention the way that it is easy to access, cyclic voltammograph in the form of potentiostat and specific programs is a must in most research and development labs. **Table 2** summarizes the use of CV in other areas besides chemistry.

4. Conclusive remarks

Cyclic voltammetry has long provided essential information in a number of areas and obviously will continue to facilitate research work as an analytical tool. The technique gives researchers an insight into the systems they are working on. For chemistry, the technique has shed the light on the mechanism of electron transfer reactions as well as their quantitative and synthetic relevance. The way that the electron transfer responses to the parameters of the surrounding environment leads scientists to be able to follow the behavior of target substances and therefore supports their characterizations. With smaller and cheaper instruments in addition

Areas	Reference for typical example
Paint	[70]
Medical	[71]
Engineering	[72]
Pharmaceutical sciences	[73–75]
Polymer and polymer films	[76]
Material sciences • microporous materials • semiconductors	[77]
Surface science and interface	[78]
Corrosion	[79]
Energy	[80]
Battery	[81]
Fuel	[82]
Fuel cell and microbial fuel cell	[41]
Processing and production	[83]
Flow detection	[84]
Food • wine product • antioxidant • preservation	[85]
Environment	[86]
Catalysis	[87]
Forensics	[88]

Table 2.
Summary of the main uses of CV in other areas besides chemistry.

to the ease the ease to operate, the technique can be available anywhere to serve all types of needs. The given examples are just only a handful of its usefulness in applications to initiate the understanding and further search. It is hoped that this review can form fundamental database to be implemented for further uses in the areas of individual interest. The opportunity is still there to investigate as well as to develop beneficial innovations. The more newly things are discovered and invented, the greater role the CV data play in feeding the world with much more meaningful scientific prosperity.

DOI: http://dx.doi.org/10.5772/intechopen.83451

Author details

Pipat Chooto
Department of Chemistry, Faculty of Science, Prince of Songkla University, Hatyai, Songkhla, Thailand

*Address all correspondence to: pipat.c@psu.ac.th

References

[1] Bocasly AB. Cyclic voltammetry, electrochemical technique. In: Characterization of Materials. 2nd ed. Wiley; 2012. DOI: 10.1002/0471266965.com050.pub2

[2] Santos CMM, Garcia MBQ, Silva AMS, Santus R, Morlière P, Fernandes E. Electrochemical characterization of bioactive hydroxyxanthones by cyclic voltammetry. Tetrahedron Letters. 2013;**54**:85-90

[3] Auguste AFT, Quand-Meme GC, Ollo K, Berté Mohamed B, Sahi placide S, Ibrahima S, et al. Electrochemical oxidation of amoxicillin in its commercial formulation on thermally prepared RuO_2/Ti. Journal of Electrochemical Science and Technology. 2016;7:82-89

[4] Quand-Même GC, Auguste AFT, Hélène LEM, Ibrahima S, Ouattara Lassine O. Electrochemical oxidation of amoxicillin in its pharmaceutical formulation at boron doped diamond (BDD) electrode. Journal of Electrochemical Science and Engineering. 2015;**5**:129-143. DOI: 10.5599/jese.186

[5] Orata D, Amir Y, Nineza C, Mukabi M. Electrochemical characterization of amoxycillin, a broad spectrum antibiotic on a bentonite host matrix, using cyclic voltammetry. IOSR Journal of Applied Chemistry. 2015;7:50-58

[6] Remes A, Ihos M, Manea F. Electrochemical characterization of some electrode materials for pharmaceutically active compounds degradation. Chemical Bulletin of Politehnica University of Timisoara. 2010;**55**(2):152-155

[7] Bott AW. Electrochemical techniques for the characterization of redox polymers. Current Separations. 2001;**19**:3

[8] Aoki A, Miyashita T. Electrochemical characterization of redox polymer Langmuir–Blodgett films containing ferrocene derivatives. Macromolecules. 1996;**29**:4662-4667

[9] Kuralay F, Erdem A, Abacı S, Ozyörük H. Electrochemical characterization of redox polymer modified electrode developed for monitoring of adenine. Colloids and Surfaces. B, Biointerfaces. 2013;**105**:1-6. DOI: 10.1016/j.colsurfb.2012.12.049

[10] Brown KL, Hou X, Banks O, Krueger KA, Hinson J, Peaslee GF, et al. Characterization of tris (5-amino-1,10-phenanthroline) ruthenium(II/III) polymer films using cyclic voltammetry and Rutherford backscattering spectrometry. International Journal of Chemistry. 2011;**3**:12-19

[11] Zoppi RA, De Paoli M-A. Electrochemical characterization of polypyrrole and poly pyrrole/EPDM rubber blends by cyclic voltammetry and impedancimetry. Journal of the Brazilian Chemical Society. 1994;**5**:197-201

[12] Walton DJ, Hall CE, Chyla A. Characterization of poly(pyrroles) by cyclic voltammetry. The Analyst. 1992;**117**:1305-1311

[13] Yeu T, Yin K-M, Carbajal J, White RE. Electrochemical characterization of electronically conductive polypyrrole on cyclic voltammograms. Journal of The Electrochemical Society. 1991;**138**:2869-2877

[14] Sharma PK, Gupta G, Singh VV, Tripathi BK, Pandey P, Boopathi M, et al. Synthesis and characterization of polypyrrole by cyclic voltammetry at different scan rate and its use in electrochemical reduction of the simulant of nerve agents. Synthetic Metals. 2010;**160**:2631-2637

[15] Pruneanu S, Veress E, Marian I, Oniciu L. Characterization of polyaniline by cyclic voltammetry and UV-vis absorption spectroscopy. Journal of Materials Science. 1999;**34**:2733-2739

[16] Innuphat C, Chooto P. Determination of trace levels of Cd(II) in tap water samples by anodic stripping voltammetry with an electrografted boron-doped diamond electrode. ScienceAsia. 2017;**43**:33-41

[17] Ramsson ES, Cholger D, Dionise A, Poirier N, Andrus A, Curtiss R. Characterization of fast-scan cyclic voltammetric electrodes using paraffin as an effective sealant with in vitro and in vivo applications. PLoS One. 2015;**10**(10):e0141340. DOI: 10.1371/journal.pone.0141340

[18] Cabral MF, Pedrosa VA, Suffredini HB, Moreno RMB, Mattoso LHC, Goncalves PS, et al. Characterization of conductive natural rubber by cyclic voltammetry and electrochemical impedance spectroscopy. Zaštita Materijala. 2006;**47**:41-45

[19] Killa HM, Mercer EE, Philp RH. Applications of cyclic voltammetry in the characterization of complexes at low ligand concentration. Analytical Chemistry. 1984;**56**:2401-2405

[20] Chuaysong R, Chooto P, Pakawatchai C: Electrochemical properties of copper(I) halides and substituted thiourea complexes. ScienceAsia. 2008;**34**:440-442. DOI: 10.2306/scienceasia1513-1874.2008.34.440

[21] Knittl ET, Rusakov DA, Korotkova EI, Dorozhko EV, Voronova OA, Plotnikov EV, et al. Characterization of a novel dioxomolybdenum complex by cyclic voltammetry. Analytical Letters. 2015;**48**:2369-2379

[22] Abdou SN, Faheim AA, Alaghaz A-NMA. Synthesis, spectral characterization, cyclic voltammetry, molecular modeling and catalytic activity of sulfa-drug divalent metal complexes. Current Synthetic and Systems Biology. 2013;**2**:112. DOI: 10.4172/2332-0737.1000112

[23] Mandadapu G. Characterization of Organic Semiconductors Using Cyclic Voltammetry. Kingsville: Texas A&M University, ProQuest Dissertations Publishing; 2015. p. 10020622

[24] Sobrova P, Ryvolova M, Hubalek J, Adam V, Kizek R. Voltammetry as a tool for characterization of CdTe quantum dots. International Journal of Molecular Sciences. 2013;**14**:13497-13510. DOI: 10.3390/ijms140713497

[25] Zeng YN, Zheng N, Osborne PG, Li YZ, WB C, Wen MJ. Cyclic voltammetry characterization of metal complex imprinted polymer. Journal of Molecular Recognition. 2002;**15**: 204-208. DOI: 10.1002/jmr.578

[26] Hu Z, Troyk P, Cogan S. Comprehensive cyclic voltammetry characterization of AIROF microelectrodes. Conference Proceedings: Annual International Conference of the IEEE Engineering in Medicine and Biology Society. 2005;**5**:5246-5249

[27] Nosek M, Batys P, Skoczek M, Weronski P. Cyclic voltammetry characterization of microparticle monolayers. Electrochimica Acta. 2014;**133**:241-246

[28] Rotha C, Martz N, Hahn F, Léger J-M, Lamy C, Fuess H. Characterization of differently synthesized Pt-Ru fuel cell catalysts by cyclic voltammetry, FTIR spectroscopy, and in single cells. Journal of The Electrochemical Society. 2002;**149**:E433-E439. DOI: 10.1149/1.1511191

[29] Li X, Dong H, Zhang A, Wei Y. Electrochemical impedance and cyclic voltammetry characterization of a

metal hydride electrode in alkaline electrolytes. Journal of Alloys and Compounds. 2006;**426**:93-96

[30] Mabbott GA. An introduction to cyclic voltammetry. Journal of Chemical Education. 1983;**60**:697-701

[31] Bard AJ, Faulkner LR. Electrochemical Methods: Fundamentals and Applications. 2nd ed. New York: John Wiley & Sons; 2001

[32] Bott AW. Characterization of chemical reactions coupled to electron transfer reactions using cyclic voltammetry. Current Separations. 1999;**18**:9-16

[33] He R-X, Da-Wei Zha D-W. Cyclic voltammetry and voltabsorptometry studies of redox mechanism of lumazine. Journal of Electroanalytical Chemistry. 2017;**791**:103-108

[34] Yazdani J, Norouzi P, Nematollahi D, Ganjali RM. Online monitoring of electrochemical synthesis of 4-nitrocatechol using fast Fourier transform continuous cyclic voltammetry (FFTCCV) in flow system. Electrochimica Acta. 2018;**259**: 694-701

[35] Shayeh JS, Sefidbakht Y, Siadat SOR, Niknam K. Continuous fast Fourier transforms cyclic voltammetry as a new approach for investigation of skim milk k-casein proteolysis, a comparative study. International Journal of Biological Macromolecules. 2017;**103**:972-977

[36] Uchida Y, Kätelhön E, Compton RG. Cyclic voltammetry with non-triangular waveforms: Electrochemically irreversible and quasi-reversible systems. Journal of Electroanalytical Chemistry. 2018;**810**:135-144

[37] Montella C. Further investigation of the equivalence of staircase and linear scan voltammograms. III-averaged-current staircase voltammetry applied to electrochemical reactions involving adsorbed species. Journal of Electroanalytical Chemistry. 2018;**808**:348-361

[38] Shbeh M, Yerokhin A, Goodall R. Cyclic voltammetry study of PEO processing of porous Ti and resulting coatings. Applied Surface Science. 2018, In press

[39] Fazary AE, Ju Y-H, Fawy KF, Al-Shihri AS, Bani-Fwaz MZ, Alfaifi MY, et al. Nicotine—metal ion interactions in solutions: Potentiometric, cyclic voltammetry investigations and quantum chemical calculations. The Journal of Chemical Thermodynamics. 2017;**112**:283-292

[40] Hsieh M-T, Whang T-J. Mechanistic investigation on the electropolymerization of phenol red by cyclic voltammetry and the catalytic reactions toward acetaminophen and dopamine using poly(phenol red)-modified GCE. Journal of Electroanalytical Chemistry. 2017;**795**:130-140

[41] Cheshideh H, Nasirpouri F. Cyclic voltammetry deposition of nickel nanoparticles on TiO_2 nanotubes and their enhanced properties for electro-oxidation of methanol. Journal of Electroanalytical Chemistry. 2017;**797**:121-133

[42] Gürsu H, Gençten M, Şahin Y. One-step electrochemical preparation of graphene-coated pencil graphite electrodes by cyclic voltammetry and their application in vanadium redox batteries. Electrochimica Acta. 2017;**243**:239-249

[43] Semenova D, Zubov A, Silina YE, Micheli L, Gernaey KV. Mechanistic modeling of cyclic voltammetry: A helpful tool for understanding biosensor principles and supporting design optimization.

Sensors and Actuators B: Chemical. 2018;**259**:945-955

[44] Ji D, Liu L, Li S, Chen C, Lu Y, Wu J, et al. Smartphone-based cyclic voltammetry system with graphene modified screen printed electrodes for glucose detection. Biosensors & Bioelectronics. 2017;**98**:449-456

[45] Chen J, Chen C. A new data analysis method to determine pesticide concentrations by cyclic voltammetry. Measurement. 2013;**46**:1828-1833

[46] Mofidi Z, Norouzi P, Seidi S, Ganjali MR. Determination of diclofenac using electromembrane extraction coupled with stripping FFT continuous cyclic voltammetry. Analytica Chimica Acta. 2017;**972**:38-45. DOI: 10.1016/j.aca.2017.04.011

[47] Smajdor J, Piech R, Ławrywianiec M, Paczosa-Bator B. Glassy carbon electrode modified with carbon black for sensitive estradiol determination by means of voltammetry and flow injection analysis with amperometric detection. Analytical Biochemistry. 2018;**544**:7-12

[48] Masek A, Chrzescijanska E, Zaborski M. Characteristics of curcumin using cyclic voltammetry, UV-vis, fluorescence and thermogravimetric analysis. Electrochimica Acta. 2013;**107**:441-447

[49] Phongarthit K, Wongnawa M, Chooto P, Wongnawa S. Spectrophotometric study of complex formation between curcumin and Cr(III) ion: A case of heavily overlapping absorption peaks. Journal of Solution Chemistry. 2016;**45**:1468-1478. DOI: 10.1007/s10953-016-0512-8

[50] Heydari H, Gholivand MB, Abdolmaleki A. Cyclic voltammetry deposition of copper nanostructure on MWCNTs modified pencil graphite electrode: An ultra-sensitive hydrazine sensor. Materials Science and Engineering: C. 2016;**66**:16-24

[51] Ammar HB, Brahim MB, Abdelhédi R, Samet Y. Boron doped diamond sensor for sensitive determination of metronidazole: Mechanistic and analytical study by cyclic voltammetry and square wave voltammetry. Materials Science and Engineering: C. 2016;**59**:604-610

[52] Kong D, Kong W, Khan ZUH, Wan P, Chen Y, Yang M. Determination of thiol content in fossil fuel by cyclic voltammetry using in situ bismuth film electrode. Fuel. 2016;**182**:266-271

[53] Freitas P, Dias LG, Peres AM, Castro LM, Veloso ACA. Determination of 2,4,6-trichloroanisole by cyclic voltammetry. Procedia Engineering. 2012;**47**:1125-1128

[54] Makhotkina O, Paul A, Kilmartin PA. The phenolic composition of Sauvignon blanc juice profiled by cyclic voltammetry. Electrochimica Acta. 2012;**83**:188-195

[55] Makhotkina O, Kilmartin PA. The use of cyclic voltammetry for wine analysis: Determination of polyphenols and free sulfur dioxide. Analytica Chimica Acta. 2010;**668**:155-165

[56] Ni W, Liu S, Fei Y, He Y, Ma X, Lu L, et al. The determination of 1-methylimidazole in room temperature ionic liquids based on imidazolium cation by cyclic voltammetry. Journal of Electroanalytical Chemistry. 2017;**787**:139-144

[57] Cao X, Xu H, Ding S, Ye Y, Ge X, Yu L. Electrochemical determination of sulfide in fruits using alizarin-reduced graphene oxide nanosheets modified electrode. Food Chemistry. 2016;**194**:1224-1229.

[58] Cui L, Hu J, Li C, Wang C, Zhang C. An electrochemical biosensor based

on the enhanced quasi-reversible redox signal of prussian blue generated by self-sacrificial label of iron metal-organic framework. Biosensors and Bioelectronics. 2018;**122**:168-174

[59] Chooto C. Modified electrodes for determining trace metal ions. In: Stoytcheva M, Zlatev R, editors. Application of the Voltammetry. Rijeka, Croatia: InTech; 2017

[60] Lu Y, Liang X, Niyungeko C, Zhou J, Xu J, Tian G. A review of the identification and detection of heavy metal ions in the environment by voltammetry. Talanta. 2018;**178**: 324-338

[61] March G, Nguyen TD, Piro B. Modified electrodes used for electrochemical detection of metal ions in environmental analysis. Biosensors. 2015;**5**:241-275. DOI: 10.3390/bios5020241

[62] Santos CMM, Garcia M, BQG SAMS, Santus R, Fernandes E. Electrochemical analysis of nanostructured iron oxides using cyclic voltammetry and scanning electrochemical microscopy. Electrochimica Acta. 2016;**222**:1326-1334

[63] Ding W, Bonk A, Gussone J, Bauer T. Cyclic voltammetry for monitoring corrosive impurities in molten chlorides for thermal energy storage. Energy Procedia. 2017;**135**:82-91

[64] Tran HV, Huynh CD, Tran HV, Piro B. Cyclic voltammetry, square wave voltammetry, electrochemical impedance spectroscopy and colorimetric method for hydrogen peroxide detection based on chitosan/silver nanocomposite. Arabian Journal of Chemistry. 2016, In press

[65] Jara-Palacios MJ, Escudero-Gilete ML, Hernández-Hierro JM, Heredia FJ, Hernanz D. Cyclic voltammetry to evaluate the antioxidant potential in winemaking by-products. Talanta. 2017;**165**:211-215

[66] Roldan-Cruz C, Vernon-Carter EJ, Alvarez-Ramirez J. Assessing the stability of tween 80-based O/W emulsions with cyclic voltammetry and electrical impedance spectroscopy. Colloids and Surfaces A: Physicochemical and Engineering Aspects. 2016;**511**:145-152

[67] Lucia LL, Sbarcea G, Ioan VBIV. Cyclic voltammetry for energy level estimation of organic materials. UPB Scientific Bulletin, Series B. 2013;**75**:111-118

[68] Kang J, Kim T, Tak Y, Lee J-H, Yoon J. Cyclic voltammetry for monitoring bacterial attachment and biofilm formation. Journal of Industrial and Engineering Chemistry. 2012;**18**: 800-807

[69] Bhattacharyya R, Tudu B, Das SC, Bhattacharyya N, Bandyopadhyay R, Pramanik P. Classification of black tea liquor using cyclic voltammetry. Journal of Food Engineering. 2012;**109**:120-126

[70] Bowman L, Spencer D, Muntele C, Muntele I, Ila D. Cyclic voltammetry and RBS study of paint components. Nuclear Instruments and Methods in Physics Research Section B: Beam Interactions with Materials and Atoms. 2007;**261**:557-560

[71] de Abreua FC, Ferraz PAL, Goulart MOF. Some applications of electrochemistry in biomedical chemistry. Emphasis on the correlation of electrochemical and bioactive properties. Journal of the Brazilian Chemical Society. 2002;**13**:19-35. DOI: 10.1590/S0103-50532002000100004

[72] Jittiarporn P, Sikong L, Kooptarnond K, Taweepreda W, Chooto P, Khangkhamano M. Synthesis of h-MoO_3 and $(NH_4)_2Mo_4O_{13}$ using precipitation method at various

pH values and their photochromic properties. Applied Mechanics and Materials. 2016;**835**:34-41

[73] Ozkan SA, Kauffmann J-M, Zuman P. Electroanalysis in Biomedical and Pharmaceutical Sciences: Voltammetry, Amperometry, Biosensors, Applications. Berlin Heidelberg: Springer-Verlag; 2015

[74] Mekassa B, Tessema M, Chandravanshi BS, Baker PGL, Muya FN. Sensitive electrochemical determination of epinephrine at poly(L-aspartic acid)/electro-chemically reduced graphene oxide modified electrode by square wave voltammetry in pharmaceutics. Journal of Electroanalytical Chemistry. 2017;**807**:145-153

[75] El-Hady MNA, Gomaa EA, Al-Harazie AG. Cyclic voltammetry of bulk and nano $CdCl_2$ with ceftazidime drug and some DFT calculations. Journal of Molecular Liquids. 2018, In press

[76] Do J-S, Chang Y-H, Tsai M-L. Highly sensitive amperometric creatinine biosensor based on creatinine deiminase/Nafion®-nanostructured polyaniline composite sensing film prepared with cyclic voltammetry. Materials Chemistry and Physics. 2018;**219**:1-12

[77] Mei B-A, Pilon L. Three-dimensional cyclic voltammetry simulations of EDLC electrodes made of ordered carbon spheres. Electrochimica Acta. 2017;**255**:168-178

[78] Viada BN, Juárez AV, Gómez EMP, Fernández MA, Yudi LM. Determination of the critical micellar concentration of perfluorinated surfactants by cyclic voltammetry at liquid/liquid interfaces. Electrochimica Acta. 2018;**263**:499-507

[79] Chooto P, Manaboot S. Electrochemical, spectrochemical and quantum chemical studies on dimedone as corrosion inhibitor for copper in acetonitrile. Journal of Scientific Research and Reports. 2017;**15**: 1-13

[80] Inoue M, Cedeño R. A dimeric tris(2,2′-bipyridine)ruthenium(II) system: Emission spectrum and cyclic voltammogram. Inorganica Chimica Acta. 1988;**145**:117-120

[81] Hansen S, Quiroga-González E, Carstensen J, Föll H. Size-dependent cyclic voltammetry study of silicon microwire anode for lithium ion batteries. Electrochimica Acta. 2016;**217**:283-291

[82] Nechaeva D, Shishov A, Ermakov S, Bulatov A. A paper-based analytical device for the determination of hydrogen sulfide in fuel oils based on headspace liquid-phase microextraction and cyclic voltammetry. Talanta. 2018;**183**:290-296

[83] Pouri SR, Manic M, Phongikaroon S. A novel framework for intelligent signal detection via artificial neural networks for cyclic voltammetry in pyroprocessing technology. Annals of Nuclear Energy. 2018;**111**:242-254

[84] Jiao Y, Kilmartin PA, Fan M, Quek SY. Assessment of phenolic contributors to antioxidant activity of new kiwifruit cultivars using cyclic voltammetry combined with HPLC. Food Chemistry. 2018;**268**:77-85

[85] Schneider M, Türke A, Fischer W-J, Kilmartin PA. Determination of the wine preservative sulphur dioxide with cyclic voltammetry using inkjet printed electrodes. Food Chemistry. 2014;**159**:428-432

[86] Massah J, Vakilian KA. An intelligent portable biosensor for fast and accurate nitrate determination using cyclic voltammetry. Biosystems Engineering. 2018, In press

[87] Surendra BS. Green engineered synthesis of Ag-doped $CuFe_2O_4$: Characterization, cyclic voltammetry and photocatalytic studies. Journal of Science: Advanced Materials and Devices. 2018;**3**:44-50

[88] Üzer A, Sağlam Ş, Tekdemir Y, Ustamehmetoğlu B, Apak R. Determination of nitroaromatic and nitramine type energetic materials in synthetic and real mixtures by cyclic voltammetry. Talanta. 2013;**115**:768-778

Chapter 2

Tips of Voltammetry

Koichi Jeremiah Aoki and Jingyuan Chen

Abstract

Theories of cyclic voltammetry, AC-impedance techniques, and the double-layer capacitive currents are described concisely to touch their principles. Applications of the theory to experimental data do not always lead to reasonable interpretation consistent with other techniques. Several tips are presented not only in the experimental viewpoint but also in a perspective of the data analysis. Most of them are devoted to cyclic voltammetry. They include shape of voltammograms, information from peak currents and peak potentials, criteria of diffusion and adsorption controls, the static and the dynamic numbers of electrons, handling of reference and counter electrodes, usage of AC impedance, concepts of heterogeneous charge-transfer rates, and combination with data by scanning probe microscope. They belong partially to recommendation and prohibition.

Keywords: current-voltage curves, cyclic voltammetry, AC impedance, double-layer capacitance, diffusion control, surface wave

1. Introduction

Voltammetry is an electrochemical technique for current-voltage curves, from which electrode reactions at electrode-solution interfaces can be interpreted. Since current-voltage curves, called voltammograms, include sensitive properties of solution compositions and electrode materials, their analysis provides not only chemical structures and reaction mechanisms on a scientific basis but also electrochemical manufacture on an industrial basis. The voltammograms vary largely with measurement time except for steady-state measurements, and so it is important to pay attention to time variables. Voltage is a controlling variable in conventional voltammetry, and the current is a measured one detected as a function of applied voltage at a given time.

The equipment for voltammetry is composed of electrodes, solution, and electric instruments for voltage control. Electrodes and electric instruments are keys of voltammetry. Three kinds of electrodes are desired to be prepared: a working electrode, a counter one, and a reference one. The three will be addressed below.

Let us consider a simple experiment in which two electrodes are inserted into a salt-included aqueous solution. When a constant current is applied to the two electrodes, reaction $2H^{+} + 2e^{-} \rightarrow H_2$ may occur at one electrode, and reaction $2OH^{-} \rightarrow H_2O_2 + 2e^{-}$ occurs at the other. The current is the time variation of the electric charge, and hence it is a kind of reaction rate at the electrode. Since the applied current is a sum of the two reaction rates, one being in the positive direction and the other being in the negative, it cannot be attributed to either reaction rate. A technique of attributing the reactions is to use an electrode with such large area that an uninteresting reaction rate may not become a rate-determining step.

This electrode is called a counter electrode. The current density at the counter electrode does not specifically represent any reaction rate. In contrast, the current density at the electrode with a small area stands for the interesting reaction rate. This electrode is called a working electrode. It is the potential difference, i.e., voltage, at the working electrode and in the solution that brings about the electrode reaction. However, the potential in the solution cannot be controlled with the working electrode or the counter one. The control can be made by mounting another electrode, called a reference electrode, which keeps the voltage between an electrode and a solution to be constant. However, the constant value cannot be measured because of the difference in phases. A conventionally employed reference electrode is silver-silver chloride (Ag-AgCl) in high concentrated KCl aqueous solution.

An electric instrument of operating the three electrodes is a potentiostat. It has three electric terminals: one being a voltage follower for the reference electrode without current, the second being a current feeder at the counter electrode, and the third being at the working electrode through which the current is converted to a voltage for monitoring. A controlled voltage is applied between the working electrode and the reference one. These functionalities can readily be attained with combinations of operational amplifiers. A drawback of usage of operational amplifiers is a delay of responses, which restricts current responses to the order of milliseconds or 10 kHz frequency.

Voltammetry includes various types—linear sweep, cyclic, square wave, stripping, alternating current (AC), pulse, steady-state microelectrode, and hydrodynamic voltammetry—depending on a mode of the potential control. The most frequently used technique is cyclic voltammetry (CV) on a time scale of seconds. In contrast, currently used voltammetry at time as short as milliseconds is AC voltammetry. We describe here the theory and tips for practical use of mainly the two types of voltammetry.

2. Theory

The theory of voltammetry is to obtain expressions for voltammograms on a given time scale or for those at a given voltage. First of all, it is necessary to specify rate-determining steps of voltammograms. There are three types of rate-determining steps under the conventional conditions: diffusion of redox species in solution near an electrode, adsorption on an electrode, and charging processes at the double layer (DL). Electric field-driven mass transport, called electric migration, belongs to rare experimental conditions, and hence it is excluded in this review. When a redox species in solution is consumed or generated at an electrode, it is supplied to or departed from the electrode by diffusion unless solution is stirred. When it is accumulated on the electrode, the change in the accumulated charge by the redox reaction provides the current. Whenever electrode voltage is varied with the time, the charging or discharging of the DL capacitor causes current. Therefore, the three steps are frequently involved in electrochemical measurements.

A mass transport problem on voltammetry is briefly described here. The redox species is assumed to be transported by one-directional (x) diffusion owing to heterogeneous electrode reactions. Then, the flux is given by $f = -D(\partial c/\partial x)$, where c and D are the concentration and the diffusion coefficient of the redox species, respectively. Redox species in solution causes some kinds of chemical reaction through chemical reaction rates, $h(c, t)$. Then the reaction rate is the sum of the diffusional flux and the chemical reaction rate, $\partial c/\partial t = -\partial f/\partial x - h(c, t)$. Here the equation for $h = 0$ is called an equation of continuum. Eliminating f with the above equation on the assumption of a constant value of D yields $\partial c/\partial t = D(\partial^2 c/\partial x^2) - h(c, t)$. This is an equation for diffusion-chemical kinetics. The expression at $h = 0$ is the diffusion equation. A boundary condition with electrochemical significance is the control of c at the electrode surface with a given electrode potential. If the redox reaction occurs in equilibrium with the one-electron transfer at the electrode, the Nernst equation for the concentrations of the oxidized species, c_o, and the reduced one, c_r, holds.

$$(c_o/c_r)_{x=0} = \exp\left[F(E\text{–}E^o)/RT\right] \quad (1)$$

where E^o is the formal potential. If there is no adsorption, the zero-flux condition in the absence of accumulation is valid:

$$D_o(\partial c_o/\partial x)_{x=0} + D_r(\partial c_r/\partial x)_{x=0} = 0 \quad (2)$$

The other conditions are concentrations in the bulk ($x \to \propto$) and the initial conditions.

2.1 Diffusion-controlled current

If the mass transport is controlled only by x-directional diffusion, c_r and c_o are given by the diffusion equations, $\partial c/\partial t = D(\partial^2 c/\partial t^2)$ for $c = c_r$ or c_o. An electrochemically significant quantity is not concentration in any x and t, but a relation between the surface concentrations and the current (the flux at $x = 0$). On the assumption of $D_o = D_r = D$, of the initial and boundary conditions, $(c_r)_{t=0} = c^*$, $(c_o)_{t=0} = 0$, and $(c_r)_{x=\infty} = c^*$, $(c_o)_{x=\infty} = 0$, a solution of the initial-boundary problem is given by [1].

$$(c_o)_{x=0} = (\pi D)^{-\frac{1}{2}} \int_0^t [j(u)/F](t-u)^{-\frac{1}{2}} du \quad (3)$$

where j is the current density. The common value of the diffusion coefficients yields $c_o + c_r = c^*$ for any x and t. Inserting this relation and Eq. (3) into the Nernst equation, $(c_o)_{x=0} = c^*/[1 + \exp[-F(E - E^o)/RT]]$, we obtain the integral equation for j as a function of t or E.

2.1.1 *Linear sweep voltammetry by diffusion*

When the voltage is linearly swept with the time at a given voltage scan rate, v, from the initial potential E_{in}, Eq. (3) through the combination with the Nernst equation becomes

$$\sqrt{\pi D}c^*F/\left[1 + \exp\left(-\frac{F(E_{in} + vt - E^o)}{RT}\right)\right] = \int_0^t j(u)(t-u)^{-\frac{1}{2}} du \quad (4)$$

The above Abel's integral equation can be solved by Laplace transformation. When the time variation is altered to the voltage variation through $E = E_{in} + vt$, the current density is expressed as

$$j = \frac{c^*F^{3/2}}{4}\sqrt{\frac{Dv}{\pi RT}} \int_{\zeta_i}^{\zeta} (\zeta - u)^{-1/2} \text{sech}^2\left[\frac{\zeta(u)}{2}\right] du \quad (5)$$

where $\zeta = (E - E^o)F/RT$ and $\zeta_i = (E_{in} - E^o)F/RT$. Evaluation of the integral has to resort to numerical computation. Current at any voltage should be proportional to

$v^{1/2}$, as can be seen in Eq. (5). The voltammogram for $v > 0$ rises up from E_o, takes a peak, and then deceases gradually with the voltage. The decrease in the current is obviously ascribed to relaxation by diffusion. The peak current density is expressed by

$$j_p = 0.446c^* F^{3/2}(Dv/RT)^{1/2} \tag{6}$$

at $E_p = E^o + 0.029$ V at 25°C, where 0.446 comes from the numerical calculation of the integral of Eq. (5).

Practical voltage-scan voltammetry is not simply linear sweep but cyclic voltammetry (CV), at which applied voltage is reversed at a given voltage in the opposite direction. The theoretical evaluation of the voltammogram should be at first represented in the integral form with the time variation and then express the time as the voltage. One of the features of the diffusion-controlled cyclic voltammograms is the difference between the anodic peak potential and the cathodic one, ΔE_p (in **Figure 1**), of which value is 59 mV at 25°C.

2.1.2 AC voltammetry by diffusion

AC voltammetry can be performed when the time variation of voltage is given by $E = E_{dc} + V_0 e^{i\omega t}$, where ω is the frequency of applied AC voltage, i is the imaginary unit, V_0 is its voltage amplitude, and E_{dc} is the DC voltage. A conventional value of V_0 is 10 mV. When this voltage form is inserted into Eq. (3) together with the Nernst equation, the AC component of the current density is represented by [2].

$$j = (1+i)\left[c^* F^2 (D\omega/2)^{1/2} V_0 \, e^{i\omega t}/RT\right] \, \mathrm{sech}^2\{(E_{dc} - E^o)/RT\}] \tag{7}$$

A voltammogram (j vs. E_{dc}) at a given frequency takes a bell shape, which is expressed by $\mathrm{sech}^2\{(E_{dc} - E^o)/RT\}$. The functional form of sech^2 is shown in **Figure 2**. The peak current appears at $E_{dc} = E^o$.

The AC-impedance technique often deals with the real impedance, Z_1, $= 1/2Y_1$ and the imaginary one, $Z_2 = -1/2Y_1$, where Y_1 is the real admittance given by

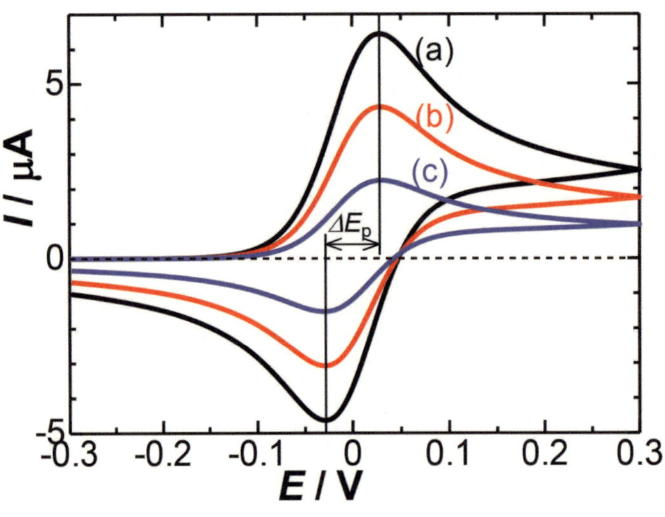

Figure 1.
Voltammograms calculated from Eq. (5) for v = (a) 180, (b) 80 and (c) 20 mV s⁻¹.

$$Y_1 = \left[c^* F^2 (D\omega/2)^{1/2}/RT\right] \operatorname{sech}^2[(E_{dc} - E^o)/RT] = Y_2 \tag{8}$$

Here Y_2 is the imaginary admittance, equal to Y_1. Since $Z_1 = -Z_2$, the Nyquist plot, i.e., $-Z_2$ vs. Z_1, is a line with the slope of unity. The term $1 + i$ in Eq. (7) has come from $(D\omega)^{1/2}$, originating from $(Di\omega)^{1/2}$. Therefore, it can be attributed to diffusion. In other words, diffusion produces the capacitive component as a delay.

2.2 Adsorption-controlled current

When the redox species with reaction $R = O + e^-$ is adsorbed on the electrode and has no influence from the redox species in the solution, the sum of the surface concentrations of R and O is a constant, Γ^*. Then the surface concentration of the oxidized species, Γ_o, is given by the Nernst equation:

$$\Gamma_o = \Gamma^*/[1 + \exp(-(E - E^o)F/RT)] \tag{9}$$

2.2.1 Linear sweep voltammetry by adsorption

The time derivative of the redox charge corresponds to the current density, $j = d(F\Gamma_o)/dt$. Application of the condition of voltage sweep, $E = E_{in} + vt$, to Eq. (9) yields.

$$j = \left(F^2\Gamma^* v/4RT\right) \operatorname{sech}^2(\zeta/2) \tag{10}$$

The voltammogram takes a bell shape (**Figure 2**), of which peak is at $E = E^o$, similar to the AC voltammogram. The current at any voltage is proportional to v. Since the negative-going scan of the voltage provides negative current values, the cyclic voltammogram should be symmetric with respect to the $I = 0$ axis. The peak current is expressed as $j_p = F^2\Gamma^* v/4RT$. The width of the wave at $j_p/2$ is 90 mV at 25°C.

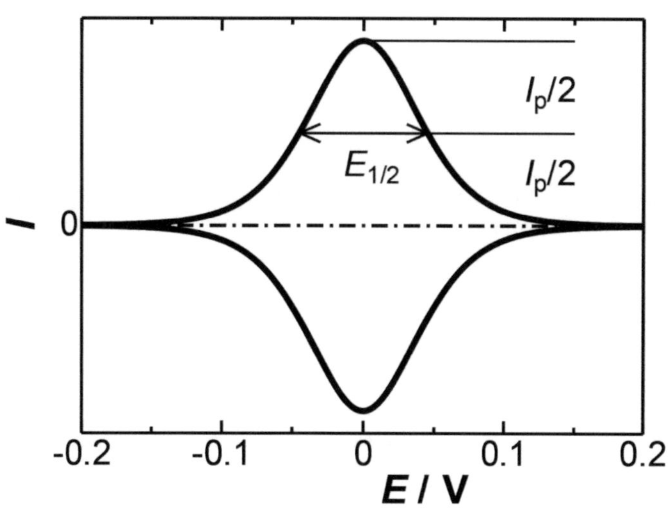

Figure 2.
Voltammogram calculated from Eq. (10).

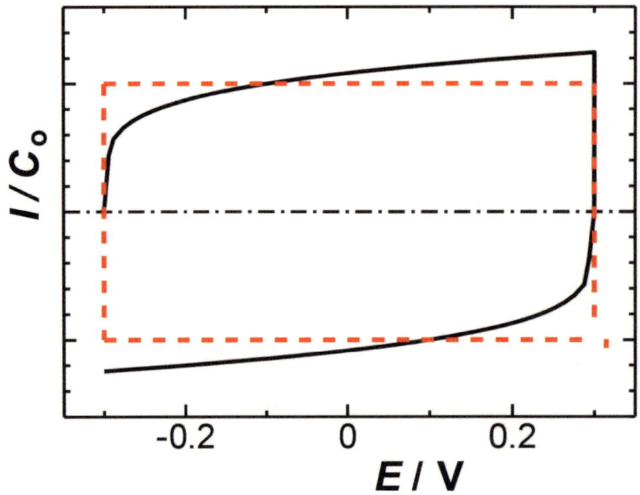

Figure 3.
Capacitive voltammograms by CV at $v = 0.5\ V\,s^{-1}$ *for (dashed lines) the ideal capacitance and for Eq. (13) (solid curves) at* $\lambda = 0.2$.

2.3 Capacitive current

Since a phase has its own free energy, contact of two phases provides a step-like gap of the free energy, of which gradient brings about infinite magnitude of force. In order to relax the infinity, local free energy varies from one phase to the other as smoothly as possible at the interface. The large variation of the energy is compensated with spontaneously generated space variations of voltage, i.e., the electric field, which works as an electric capacitor. The capacitance at solution-electrode interface causes orientation of dipoles and nonuniform distribution of ionic concentration, of which layer is called an electric double layer (DL).

When the time variation of the voltage is applied to the DL capacitance, C_d, the definitions of the capacitance ($q = C_d V$) and the current lead

$$I = \mathrm{d}(C_d V)/\mathrm{d}t = C_d(\mathrm{d}V/\mathrm{d}t) + V(\mathrm{d}C_d/\mathrm{d}t) \tag{11}$$

where C_d generally depends on the time. This dependence is significant for understanding experimentally observed capacitive currents.

2.3.1 Capacitance by AC impedance

The DL capacitance has exhibited the frequency dispersion expressed by $C_d = (C_d)_{1\mathrm{Hz}} f^{-\lambda}$, called the constant phase element [3–5] or power law [6, 7], where λ is close to 0.1. Inserting this expression and $V = V_0 e^{i\omega t}$ into Eq. (11) yields

$$I = (i + \lambda)\omega C_d V \tag{12}$$

This is a simple sum of the real part of the current and the imaginary one, indicating that the equivalent circuit should be a parallel combination of a capacitive component and a resistive one, both depending on frequency. Since the ratio, $-Z_2/Z_1$, for Eq. (12) is $1/\lambda$, the Nyquist plots have slopes less than 10 rather than infinity.

2.3.2 Capacitive current by CV

If the capacitive charge is independent of the time, the capacitive current should be $I = d(CV)/dt = C(E - E^{o})/v$. Therefore, it takes a horizontal positive ($v > 0$) and a negative line ($v < 0$), as shown in **Figure 3** (dashed lines). When the time dependence of C, i.e., $C_d = (C_d)_0 t^{-\lambda}$, is applied to Eq. (11), for the forward and the backward scans, respectively, we have

$$I = (\lambda + 1)vC_0t^{\lambda}, \text{and } I = (\lambda + 1)vC_0t^{\lambda} \quad (13)$$

The variation of CV computed from Eq. (13) (**Figure 3**, solid curves) is similar to our conventionally observed capacitive waves.

3. Tips of voltammetric analysis

Voltammograms can identify an objective species by comparing a peak potential with a table of redox potentials and furthermore determine its concentration from the peak current. Their results are, however, sometimes inconsistent with data by methods other than electrochemical techniques if one falls in some pitfalls of analytical methods of electrochemistry. For example, a peak potential is influenced by a reference electrode and solution resistance relevant to methods. Peak currents are varied complicatedly with mass transport modes as well as associated chemical reactions. Since the theory on voltammetry covers only some restricted experimental conditions, it can rarely interpret the experimental data successfully. This review is devoted to some voltammetric tips which can lead experimenters to reasonable interpretation.

3.1 Understanding outline of voltammograms

It is rare to observe a reversible voltammogram in which both oxidation and reduction waves appear in a symmetric form with respect to the potential axis at a similar peak potential, as in **Figure 1**. Frequently observed voltammograms are irreversible, i.e., either a cathodic or an anodic wave appears; a value of a cathodic peak current is quite different from the anodic one in magnitude; a cathodic peak potential is far from the anodic one. These complications are ascribed to chemical reactions and/or phase transformation after the charge-transfer reaction. A typical example is deposition of metal ions on an electrode. The complications can be interpreted by altering scan rates and reverse potentials.

A wave at a backward scan is mostly attributed to electrode reactions generated by experimenters rather than to species latently present in the solution. That is, it is artificial. It is caused either by the reaction of the wave at the forward scan or the reaction of the rising-up current just before the reverse potential. A source of the backward wave can be found by changing the reverse potentials.

Some voltammograms have more than two peaks at one-directional scan. The appearance of the two can be interpreted as a two-step sequential charge-transfer reaction. However, multiple waves appear also by combinations of chemical reactions and adsorption. The peak current and the charge for this case are quite different from the predicted ones, as will be described in Section 3.2. Change in scan rates may be helpful for interpreting the multiple waves.

3.2 Shape and values of peaks

It is possible to predict theoretically a controlling step of voltammograms from their shape (a bell type corresponding to an adsorption wave or a draw-out type corresponding to a diffusion wave). However, the shape strongly depends on chemical complications, adsorption, and surface treatment of the electrodes. When redox species in solution is partially adsorbed on an electrode, the electrode process is far from a prediction because of very high concentration in the adsorbed state. A draw-out-shaped wave can be observed even for the adsorbed control. It is important to estimate which state the reacting species takes on the electrode. Potentials representing of voltammetric features do not express a controlling step in reality although the theory does. One should pay attention to the current. The peak current controlled by diffusion with one-electron transfer is given by $I_p = 0.27\, cAv^{1/2}$ μA (c, bulk concentration mM; A, electrode area mm^2; v, potential sweep rate $mV\ s^{-1}$). The microelectrode behavior sometimes comes in view at $v < 10\ mV\ s^{-1}$, $A < 0.1\ mm^2$, so the measured current is larger than the estimated value. On the other hand, the peak current controlled by adsorption is given by $I_p = 1.6\, Av$ nA when one redox molecule is adsorbed at 1 nm^2 on the electrode. The voltammogram by adsorption often differs from the ideal bell shape due to adsorbed molecular interaction and DL capacity. Division of the area of the peak by the scan rate yields the amount of adsorbed electricity. Comparison of this with the anticipated amount of adsorption may be helpful for understanding the electrode process.

3.3 Deviation of ΔE_p from theoretical values

The peak potential difference ΔE_p between the oxidation wave and the reduction wave (**Figure 1**) has been used for a prediction of the reaction mechanism. For example, $\Delta E_p = 60$ mM suggests the diffusion-controlled current accompanied by one-electron exchange, whereas $\Delta E_p = 30$ mM infers a simultaneous reaction with two electrons. Then what would happen for 120 mV which is sometimes found? A half-electron reaction might not be accepted. Potential shift over 60 mV occurs by chemical complications. In contrast, the voltammogram by adsorbed species shows theoretically a bell shape with the width, $E_{1/2} = 90$ mV, at the half height of the peak (**Figure 2**). This value is based on the assumption of the absence of interaction among adsorbed species. However, adsorption necessarily yields such high concentrations as strong interaction.

It is necessary to pay attention to the validity of analyzing ΔE_p and $E_{1/2}$. The peak potential is the first derivative of a voltammogram. Since ΔE_p is a difference between the two peaks, it is actually the second-order derivative of the curves in the view of accuracy. In other words, the accuracy of ΔE_p is lower than that of peak current. Furthermore, peak potentials as well as $E_{1/2}$ readily vary with scan rates owing to chemical reactions and solution resistance. One should use the peak current for data analysis instead of the potentials.

3.4 Criteria of diffusion-controlled currents

Voltammograms of a number of redox species have been reported to be diffusion controlled from a relationship between I_p and $v^{1/2}$. The redox species exhibiting diffusion-controlled current is, however, limited to ferrocenyl derivatives under conventional conditions. Voltammograms even for $[Fe(CN)_6]^{3-/4-}$ and $[Ru(NH_3)_6]^{3+}$ are deviated from the diffusion control for a long-time measurement. Why have many researchers assigned voltammograms to be the diffusion-controlled step? The proportionality of I_p to $v^{1/2}$ in Eq. (6) has been confused with the linearity, $I_p = av^{1/2} + b$ ($b \neq 0$). The plot for the adsorption control ($I_p = kv$) also shows approximately a linear relation for I_p vs. $v^{1/2}$ plot in a narrow domain of v, as shown in **Figure 4B**. The opposite is true (**Figure 4A**). Therefore, it is the intercept that determines a controlling step of either the diffusion or adsorption. Some may say that the intercept can be ascribed to a capacitive current. If so, the peak current should be represented by $I_p = av^{1/2} + bv$, which exhibits neither linear relation with $v^{1/2}$ nor v.

There is a simple method of determining a controlling step either by diffusion or adsorption. Current responding to diffusion-controlled potential at a disk electrode in diameter less than 0.1 mm would become under the steady state after a few seconds [8]. Adsorption-limited current should become zero soon after the potential application. Many redox species, however, show gradual decrease in the current because reaction products generate an adsorbed layer which blocks further electrode reactions.

3.5 Plots of I_p against E_p

It is well known that currents vary not only with applied voltage but also with the time. It is not popular, however, to discuss quantitatively time dependence of CV voltammograms. Enhancing v generally increases the current and causes the peak potential to shift in the direction of the scan. A reason for the former can be interpreted as generation of large current at a shorter time (see Eqs. (6) and (10)), whereas the latter is ascribed to a delay of reaction responses as well as a voltage loss of the reaction by solution resistance. Then the voltage effective to the reaction is lower than the intended voltage, and so the observed current may be smaller than the predicted one. Although I_p is related strongly with E_p, the relationship has rarely been examined quantitatively.

A technique of analyzing the potential shift is to plot I_p against E_p, [9] as shown in **Figure 5**. If the plots on the oxidation side ($I_p > 0$) and the reduction side ($I_p < 0$) fall each on a straight line, the slope may represent conductivity. If values of both slopes are equal, the slope possibly stands for the conductivity of the solution or membrane regardless of the electrode reaction. The potential extrapolated to the zero current on each straight line should be close to the formal potential. Since this plot is simple technically, the analytical result is more reliable than at least discussion of time dependence of E_p.

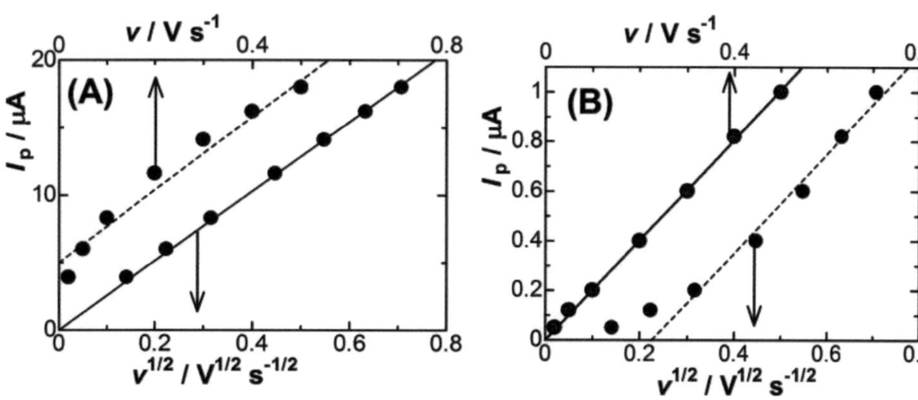

Figure 4.
Plots of I_p of (A) $K_3Fe(CN)_6$ and (B) polyaniline-coated electrode against $v^{1/2}$ and v. Both plots show approximately linear relations.

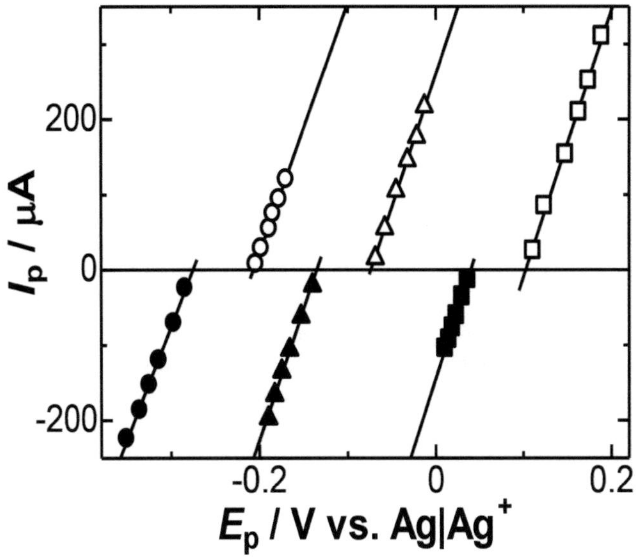

Figure 5.
Plots of I_p vs. E_p by CV of the first (circles) and the second (triangles) peak of tetracyanoquinodimethane (TCNQ), and ferrocene (squares) in 0.2 M $(CH_3)_4NPF_6$ included acetonitrile solution when scan rates were varied, where triangles were displayed by 0.4 V shift.

3.6 Meaning of $n^{3/2}$ in the equation for peak current

Most researchers have quoted the Randles-Sevcik equation, $j_p = 0.446\,(nF)^{3/2}c^*(Dv/RT)^{1/2}$, for the diffusion-controlled peak current without hesitation, where n is the electron transfer number of the reaction. According to Faraday's law, the electrolytic quantity is proportional to nc^*. Why is the peak current proportional to $n^{3/2}$ instead of n? Let us consider voltammetry of metal nanoparticles (about 25 nm in diameter) composed of 10^6 metal atoms dispersed in solution. Faraday's law predicts that the current is 10^6 times as high as the current by the one metal atom. However, Randles-Sevcik equation predicts the current further $(10^6)^{1/2} = 1000$ times as large, just by the effect of the potential scan. The order 3/2 is specific to CV. The order of n for AC current and pulse voltammetry is 2 [10]. On the other hand, the diffusion-controlled steady-state currents at a microelectrode and a rotating disk electrode are proportional to n. Comparing the differences in the order by methods, we can predict that the time variation of the voltage increases the power of n.

Let a potential width from a current-rising potential to E_p be denoted by ΔE. When an n-electron transfer reaction occurs through the Nernst equation at which F in Eq. (1) is replaced by nF, the concentration-potential curve takes the slope n times larger than that at $n = 1$ (see $c_o/c_r \cong nF(E - E^o)/RT$ near $E = E^o$ in Eq. (1)). Then we have $(\Delta E)_n = (\Delta E)_{n=1}/n$. The period of elapsing for $(\Delta E)_n$ becomes shorter by $1/n$, as if v might be larger by n times. Then v in Eq. (6) should be replaced by $(nv)^{1/2}$. Combining this result with the flux j/nF, the current becomes $n^{3/2}$ times larger than that at $n = 1$. Therefore, the factor $n^{3/2}$ results from the Nernst equation. This can be understood quantitatively by replacing F in Eq. (3) by nF. There are quite a few reactions for $n \geq 2$ both for Nernst equation and in the bulk as stable species. The term $n^{3/2}$ is valid only for a concomitant charge-transfer reaction, i.e., simultaneous occurrence n-electron transfer rather than a step-by-step transfer. Apparent two-electron transfer reactions in the bulk, for example, Cu, Fe, Zn, and Pb, cause other reactions immediately after the one-electron transfer.

3.7 Area of counter electrode

An electrochemical response is observed as a sum of the half reactions at the two electrodes. In order to extract the reaction at the working electrode, a conventional technique is to increase the area of the counter electrode so that the reaction at the counter electrode can be ignored. If the counter electrode area is increased by 20 times the area of the working electrode, the observed current represents the reaction of the working electrode with an error of 5%. Let us consider the experiment in which nanoparticles of metal are coated on a working electrode for obtaining capacitive currents or catalyst currents. Then, the actual area of the working electrode can be regarded as the area of the metal particles measured by the molecular level. Then, the area will be several thousand times the geometric area so that the observed current may represent the reaction at the counter electrode. This kind of research has frequently been found in work on supercapacitors. On the other hand, if the electrode reaction is diffusion controlled, the current is determined by the projected area of the diffusion layer. Then the current is not affected by the huge surface area of nanoparticles.

It is important to examine whether or not a reaction is controlled by at a counter electrode. A simple method is to coat nanoparticles also on the counter electrode. Then the current in the solution may become so high that the potential of the working electrode cannot be controlled. It is better to use a two-electrode system. Products at the counter electrode are possible sources of contaminants through redox cycling.

3.8 Functionality of reference electrode

The Ag-AgCl electrode is most frequently used as a reference electrode in aqueous solution because of the stable voltage at interfaces of Ag-AgCl and AgCl-KCl through fast charge-transfer steps, regardless of the magnitude of current density. The "fast step" means the absence of delay of the reaction or being in a quasi-equilibrium. The stability without delay is supported with high concentration of KCl.

When an Ag-AgCl electrode is inserted to a voltammetric solution, KCl necessarily diffuses into the solution, associated with oxygen from the reference electrode. Thus, the reference electrode is a source of contamination by salt, dichlorosilver and oxygen. It is interesting to examine how much amount a solution is contaminated by a reference electrode [9]. Time variation of ionic conductivity in the pure water was monitored immediately after a commercially available Ag-AgCl electrode was inserted into the solution. **Figure 6** shows rapid increase in the conductivity as if a solid of KCl was added to the solution. Oxygen included in the concentrated KCl may contaminate a test solution. Even the Ag-AgxO electrode, which was formed by oxidizing silver wire, increased also the conductivity, probably because the surface is in the form of silver hydroxide. As a result, no reference electrode can be used for studying salt-free electrode reactions. If neutral redox species such as ferrocene is included in a solution, the potential reference can be taken from redox potential of ferrocene.

3.9 Current flowing through the double layer

When a constant voltage is applied to the ideal capacitance C, the responding current decays in the form of $\exp(-t/RC)$, where R is a resistance in series connected with C. It has been believed that a double-layer capacitance in electrochemical system behaves as an ideal capacitor, where R is regarded as solution resistance. However, any exponential variation cannot reproduce transient currents obtained at the platinum wire electrode in KCl aqueous solution, as shown in **Figure 7**. The current decays more slowly than by $\exp(-t/RC)$, because it is approximately proportional to $1/t$. The property of non-ideal capacitance is the result of the constant phase element of the DL capacitance, as described in Section 2.3. The dependence of $1/t$ can be obtained approximately by the time derivative of $q = V_0 C_0 t^{-\lambda}$ for the voltage step V_0.

The slow decay is related with a loss of the performance of pulse voltammetry, in which diffusion-controlled currents can readily be excluded from capacitive currents. The advantage of pulse voltammetry is based on the assumption of the exponential decay of the capacitive current. Since the diffusion current with $1/t^{1/2}$ dependence is close to the $1/t$ dependence, it cannot readily be separated from the capacitive current in reality. A key of using pulse voltammetry is to take a pulse time to be so long as a textbook recommends.

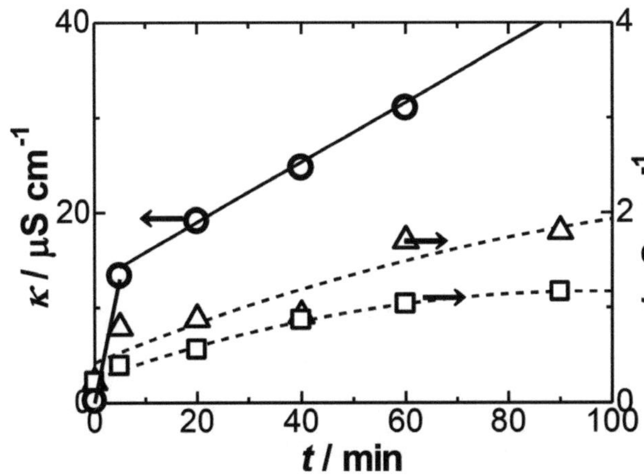

Figure 6.
Time-variation of conductivity of water into which (circles) Ag|AgCl, (triangles) Ag|Ag$_x$O, and (squares) AgCl-coated Ag wire were inserted. Conductivity measurement was under N_2 environment.

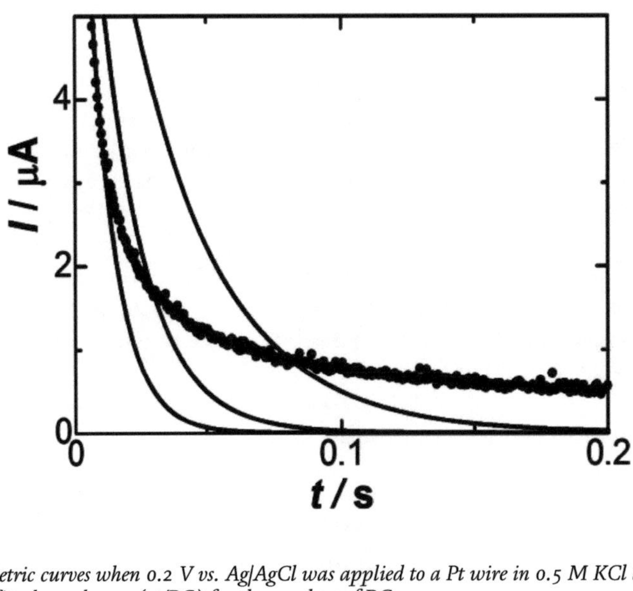

Figure 7.
*Chronoamperometric curves when 0.2 V vs. Ag|AgCl was applied to a Pt wire in 0.5 M KCl aqueous solution. Solid curves are fitted ones by exp(-t/*RC*) for three values of* RC.

High-performance potentiostats are equipped with a circuit for compensation of resistance by a positive feedback. Unfortunately, the circuit is merely useful because voltammograms depend on intensity of compensation resistances of the DL capacitance. It should work well if the DL capacitance is ideal.

3.10 Advantages of AC impedance

AC techniques have an advantage of examining time dependence at a given potential, whereas CV has a feature of finding current-voltage curves at a given time. The former shows the dynamic range from 1 Hz to 10 kHz, while the latter does conventionally from 0.01 to 1 Hz. This wide dynamic range of the AC technique is powerful for examining dynamics of electrode reactions. Analytical results by the former are often inconsistent with those by the latter, because of the difference in the time domain. The other scientific advantage of the AC technique is to get two types of independent data set, frequency variations of real components and imaginary ones by the use of a lock-in amplification. The independence allows us to operate mathematically the two data, leading to the data analysis at a level one step higher than CV. An industrial advantage is the rapid measurement, which can be applied to quality control for a number of samples. The analysis of AC impedance necessarily needs equivalent circuits of which components do not have any direction relation with electrochemical variables.

Data of the electrochemical AC impedance are represented by Nyquist (Cole-Cole) plots, that is, plots of the imaginary component (Z_2) of the impedance against the real one (Z_1), as shown in **Figure 8**. The simplest equivalent circuit for electrochemical systems is the DL capacitance C_d in series with the solution resistance R_S. The Nyquist plot for this series circuit is theoretically parallel to the vertical axis (**Figure 8A-a**), but experiments show a slope of 5 or more (**Figure 8A-b**). This behavior, called constant phase element (CPE) and the power law, has been verified for combinations of various materials and solvents [6, 7, 11, 12]. The equivalent circuit for Eq. (12) is a parallel combination of capacitance and resistance (**Figure 8B**). Even without an electrode reaction, current always includes a real component.

The equivalent circuit with the Randles type is a parallel combination of the ideal DL capacitor C_d with the ideal resistance R_{ct} representing the Butler-Volmer-type charge-transfer resistance. Practically, the Warburg impedance (the inverse of

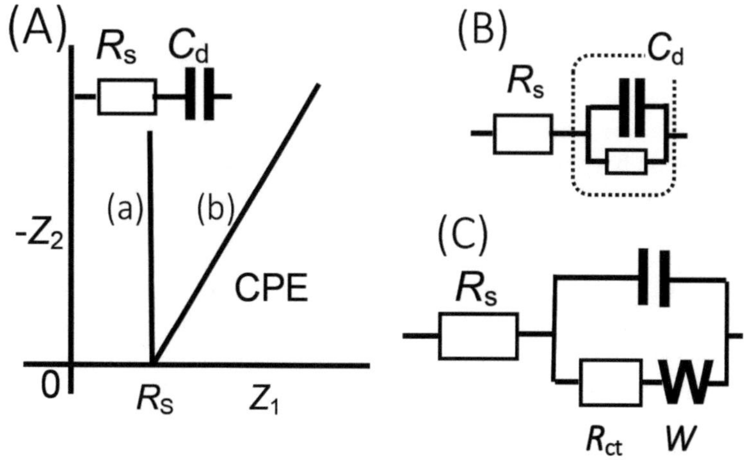

Figure 8.
(A) Nyquist plots for a RC-series circuit with ideal capacitor (a) and DL capacitor (b). (B) Equivalent circuit with the power-law of C_d. (C) Randles circuit.

Eq. (8)) due to diffusion of redox species is incorporated in a series into R_{ct} (**Figure 8C**). R_{ct} cannot be separated from the DL resistance because of the frequency dispersion. Since even the existence of R_{ct} is in question (Section 3.12), it is difficult to determine and interpret R_{ct}. The usage of a software that can analyze any Nyquist plots will provide values of R and C. Even if analyzed values are in high accuracy, researches should give them electrochemical significance.

3.11 Residual currents depending on surface treatments

Residual current varies with treatments of electrodes such as polishing of electrode surfaces and voltage applications to an extremely high domain. It can often be suppressed to yield reproducible data when the electrode is replaced by simple platinum wire or carbon rod having the same geometric area. Simple wire electrodes are quite useful especially for measurements of DL capacitance and adsorption. One of the reasons for setting off large residual current is that the insulator of confining the active area is not in close contact with the electrode, so that the solution penetrated into the gap will give rise to capacitive current and floating electrode reactions. Since the coefficient of thermal expansion of the electrode is different from that of the insulator, the residual current tends to get large with the elapse from the fabrication of the electrode. This prediction is based on experience, and there are few quantitative studies on residual currents.

Unexpected gap has been a technical problem at dropping mercury electrodes. If solution penetrates the inner wall of the glass capillary containing mercury, observed currents become irreproducible. Water repellency of the capillary tip has been known to improve the irreproducibility in order to reduce the penetration. A similar technique has been used for voltammetry at oil-water interfaces and ionic liquid-water interfaces at present.

3.12 Reversible or irreversible voltammograms

Voltammograms are said to vary with electrode reaction rates, and the rate constants have been determined from time dependence of voltammograms. The fast reaction of which rate is not rate determining has historically been called "reversible." In contrast, such a slow reaction that a peak potential varies linearly with log v is called "irreversible." A reaction between them is called "quasi-reversible." The distinction among the three has been well known since the theoretical report on the quasi-reversible reaction by Matsuda [1]. This theory is devoted to solving the diffusion equations with boundary conditions of the Butler-Volmer (BV) equation under the potential sweep. As the standard rate constant k_s in the BV equation becomes small, the peak shifts in the direction of the potential sweep from the diffusion-controlled peak. Steady-state current-potential curves in a microelectrode [13] and a rotating disk electrode also shift the potential in a similar way. According to the calculated CV voltammograms in **Figure 9**, we can present some characteristics: (i) if the oxidation wave shifts to the positive potential, the negative potential shift should also be found in the reduction wave. (ii) Both the amounts of the shift should have a linear relationship to log v. (iii) The shift should be found in iterative measurements. (iv) The peak current should be proportional to $v^{1/2}$.

The authors attempted to find a redox species with the above four behaviors. Some redox species can satisfy one of the four requirements, but do not meet the others. Most reaction rate constants have been determined from the potential shift

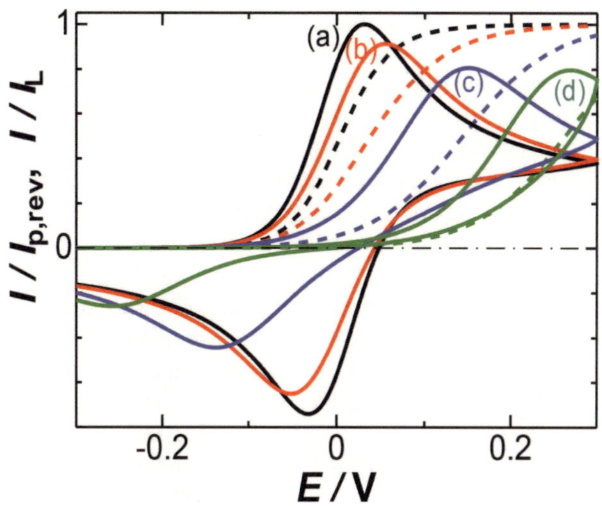

Figure 9.
CV voltammograms (solid curves) at a normally sized electrode and steady-state voltammograms (dashed curves) at a microelectrodes in 12 μm in diameter, calculated theoretically for v = 0.5 V s^{-1}, D = 0.73 × 10^{-5} cm^2 s^{-1}, ks = (a) 0.1, (b) 0.01, (c) 0.001, (d) 0.0001 cm s^{-1}. The potential shift of CV is equivalent to the wave-shift at a microelectrode through the relation, v = 0.4RTD/αFa2 (a: radius).

in a narrow time domain. They are probably caused by follow-up chemical reactions, adsorption, or DL capacitance. For example, CV peak potentials of TCNQ and benzoquinone were shifted at high scan rates, whereas their steady-state voltammograms were independent of diameters of microdisk electrodes even on the nanometer scale [14]. The shift at high scan rates should be due to the frequency dispersion of the DL capacitance, especially the parallel resistance in the DL (**Figure 8B**). Values of the heterogeneous rate constants and transfer coefficients reported so far have depended not only on the electrochemical techniques but also research groups. Furthermore, they have not been applied or extended to next developing work. These facts inspire us to examine the assumptions and validity of the BV formula.

Let us revisit the assumptions of the BV equation when an overvoltage, i.e., the difference of the applied potential from the standard electrode potential, causes the electrode reaction. The rate of the oxidation in the BV equation is assumed to have the activation energy of α times the overvoltage, while that of the reduction does that of $(1-\alpha)$ times. This assumption seems reasonable for the balance of both the oxidation and the reduction. However, the following two points should be considered. (i) Once a charge or an electron is transferred within the redox species, the molecular structure changes more slowly than the charge transfer itself occurs. The structure change causes solvation as well as motion of external ions to keep electric neutrality. These processes should be slower than the structure change. If the overvoltage can control the reaction rate, it should act on to the slowest step, which is not the genuine charge-transfer process. (ii) Since a reaction rate belongs to the probability theory, the reaction rate (dc/dt) at t is determined with the state at t rather than a state in the future. In other words, the rate of the reduction should have no relation with the oxidation state which belongs to the future state. The BV theory assumes that the α times activation energy for the oxidation is related closely with 1-α times one for the reduction. This assumption is equivalent to predicting a state at $t + \Delta t$ from state at $t + 2\Delta t$, like riding on a time machine. This question should be solved from a viewpoint of statistical physics.

3.13 Contradiction of microscopic image with electrochemical data

Development of scanning microscopes such as STM and AFM has allowed us to obtain the molecularly and atomically regulated surface images, which have been used for interpreting electrochemical data. Then the electrochemical data are expected to be discussed on a molecular scale. However, there is an essential problem of applying photographs of regularly arranged atoms on an electrode to electrochemical data, because the former and the latter include, respectively, microscopically local information and macroscopically averaged one. A STM image showing molecular patterns is information of only a part of electrode, at next parts of which no atomic images are often observed but noisy images are found. Electrochemical data should be composed of information both at a part of the electrode showing the molecular patters and at other parts showing noisy, vague images. Noisy photographs are always discarded for interpreting electrochemical data although the surfaces with noisy images also contribute electrochemical data.

An ideal experiment would be made by taking STM images over all the electrodes that provide electrochemical data and by obtaining an averaged image. However, it is not only impossible to take huge amounts of images, but the averaged image might be also noisy. It may be helpful to describe only a possibility of reflecting the STM-imaged atomic structure on the electrochemical data.

3.14 Surface wave by adsorption

Voltammograms by adsorbed redox species, called surface waves, are frequently different from a bell shape (**Figure 2**). Really observed features are the following: (i) the voltammogram does not suddenly decay after the peak, exhibiting a tail-like diffusional wave; (ii) the peak current and the amount of the electricity are proportional to the power less than the unity of v; (iii) the oxidation peak potential is different from the reduction one; (iv) the background current cannot be determined unequivocally; and (v) voltammograms depend on the starting potential. Why are experimental surface waves different from a symmetric, bell shape in **Figure 2**?

A loss of the symmetry with respect to the vertical line passing through a peak can be ascribed to the difference in interactions at the oxidized potential domain and at the reduced one. Since redox species takes extremely high concentration in the adsorbed layer, interaction is highly influenced on voltammetric form. When the left-right asymmetry is ascribed to thermodynamic interaction, it has been interpreted not only with Frumkin's interaction [15] but also Bragg-Williams-like model for the nearest neighboring interactive redox species [16]. On the other hand, most surface waves are asymmetric with respect to the voltage axis even at extremely slow scan rates. This asymmetry cannot be explained in terms of thermodynamics of intermolecular interaction, but should resort to kinetics or a delay of electrode reactions. There seems to be no delay in the electrode reaction of the monomolecular adsorption layer, different from diffusion species. The delay resembles the phenomenon of constant phase element (CPE) or frequency power law of DL capacitance, in that the redox interaction may occur two-dimensionally so that the most stable state can be attained. This behavior belongs to a cooperative phenomenon [17]. A technique of overcoming these complications is to discuss the amount of charge by evaluating the area of the voltammogram. It also includes ambiguity of eliminating background current and assuming the independence of the redox charge from the DL charge.

4. Conclusions

The simplest theories for voltammetry are limited to the rate-determining steps of diffusion of redox species and reactions of adsorbed species without interaction. Variation of scan rates as well as a reverse potential is helpful for predicting redox species and reaction mechanisms. Furthermore, the following viewpoints are useful for interpreting mechanisms:

a. comparison of values of experimental peak currents with theoretical ones, instead of discussing ΔE_p and $E_{1/2}$;

b. examining the proportionality of I_p vs. v or vs. $v^{1/2}$, i.e., zero or non-zero values of the intercept of the linearity;

c. a reference electrode and a counter electrode being a source of contamination in solution;

d. attention to very slow relaxation of DL capacitive currents;

e. inclusion of ambiguity in the equivalent circuit with the Randles type.

Author details

Koichi Jeremiah Aoki[1]* and Jingyuan Chen[2]

1 Electrochemistry Museum, Fukui, Japan

2 University of Fukui, Fukui, Japan

*Address all correspondence to: kaoki@u-fukui.ac.jp

References

[1] Matsuda H, Ayabe Y. Zur theorie der Randles-Sevčikschen kathodenstrahl-polarographie. Zeitschrift für Elektrochemie. 1955;**59**:494-503

[2] Barsoukov E, Macdonald JR. Impedance Spectroscopy. New York: Wiley; 2005. ISBN: 0-471-64749-7

[3] Lasia A. In: White RE, Conway BE, Bockris JO'M, editors. Modern Aspects of Electrochemistry. Vol. 32. New York: Kluwer Academic/Plenum Publishers; 1999. pp. 143-248

[4] Brug GJ, Van Den Eeden ALG, Sluyters-Rehbach M, Sluyters JH. The analysis of electrode impedances complicated by the presence of a constant phase element. Journal of Electroanalytical Chemistry. 1984;**176**: 275-295. DOI: 10.1016/S0022-0728(84) 80324-1

[5] Zoltowski P. On the electrical capacitance of interfaces exhibiting constant phase element behaviour. Journal of Electroanalytical Chemistry. 1998;**443**:149-154. DOI: 10.1016/ S0022-0728(97)00490-7

[6] Hou Y, Aoki KJ, Chen J, Nishiumi T. Invariance of double layer capacitance to polarized potential in halide solutions. Universal Journal of Chemistry. 2013;**1**:162-169. DOI: 10.13189/ujc.2013.010404

[7] Hou Y, Aoki KJ, Chen J, Nishiumi T. Solvent variables controlling electric double layer capacitance at metal| solution interface. Journal of Physical Chemistry C. 2014;**118**:10153-10158. DOI: 10.1021/jp5018289

[8] Zhang H, Aoki K, Chen J, Nishiumi T, Toda H, Torita E. Voltammetric determination of both concentration and diffusion coefficient by combinational use of regular and micro electrodes. Electroanalysis. 2011;**23**: 947-952. DOI: 10.1002/elan.200900603

[9] Zhang C, Aoki KJ, Chen J, Nishiumi T. Blocking of two-electron reduction of non-charged species in the absence of supporting electrolyte at nanoelectrodes. Journal of Electroanalytical Chemistry. 2013;**708**: 101-107. DOI: 10.1016/j.jelechem. 2013.09.023

[10] Smith DE. In: Bard AJ, editor. Electroanalytical Chemistry. Vol. 1. New York, N.Y.: Dekker; 1966. pp. 1-155. DOI: 10.1002/ange.19680800321

[11] Aoki KJ, Wang H, Chen J, Nishiumi T. Formation of graphite oxide nano-disks by electrochemical oxidation of HOPG. Electrochimica Acta. 2014;**130**: 381-386. DOI: 10.1016/j. electacta.2014.03.044

[12] Wang H, Aoki KJ, Chen J, Nishiumi T, Zeng Z, Ma X. Power law for frequency-dependence of double layer capacitance of graphene flakes. Journal of Electroanalytical Chemistry. 2015; **741**:114-119. DOI: 10.1016/j. jelechem.2015.01.008

[13] Aoki K. Evaluation technique of kinetic parameters for irreversible charge transfer reactions from steady-state voltammograms at microdisk electrodes. Electrochemistry Communications. 2005;**17**:523-527. DOI: 10.1016/j.elecom.2005.03.006

[14] Aoki KJ, Zhang C, Chen J, Nishiumi T. Heterogeneous reaction rate constants by steady-state microelectrode techniques and fast scan voltammetry. Journal of Electroanalytical Chemistry. 2013;**706**: 40-47. DOI: 10.1016/j. jelechem.2013.07.021

[15] Bard AJ, Faulkner LR. Electrochemical Methods: Fundamentals and Applications.

2nd ed. New York, NY: John Wiley & Sons; 2001. p. 567. ISBN: 0-471-04372-9

[16] example F, Aoki K, Chen J. Statistical thermodynamics of multi-nuclear linear complexes with mixed valence states by means of correlated-walk. Journal of Electroanalytical Chemistry. 1995;**380**:35-45. DOI: 10.1016/0022-0728(94)03603-Z

[17] Aoki KJ. Molecular interaction model for frequency-dependence of double layer capacitors. Electrochimica Acta. 2016;**188**:545-550. DOI: 10.1016/j.electacta.2015.12.049

Chapter 3

Cyclic Voltammetry of Phthalocyanines

Keiichi Sakamoto

Abstract

Phthalocyanines and their related compounds possess similar structures as porphyrins. They have been used as green to blue dyes and pigments since their discovery. In this decade, they are known to be utilized in important functional colorants for many fields such as catalyst, laser light absorbers in data storage systems, electro-charge carriers in photocopies, photo-antenna device in photosynthesis, photovoltaic cells and photosensitizers for dye-sensitized solar cells (DSSCs), and photodynamic therapy of cancer (PDT). The functions are attributed to high electron transfer abilities of phthalocyanines. Cyclic voltammograms were carried out for phthalocyanines in order to estimate their electron transfer abilities and electrochemical mechanism.

Keywords: phthalocyanines, cyclic voltammetry, electrochemistry electron transfer abilities, photosensitizers

1. Introduction

A blue-colored insoluble compound was accidently observed as a by-product at the South Metropolitan Gas Company in London during the preparation of *o*-cyanobenzamide from phthalimide and acetic acid at a high temperature by Braun and Tcherniac in 1907. The compound was later called phthalocyanine. In 1927, at the University of Fribourg, de Diesbach and von der Weid obtained stable blue material during the preparation of phthalonitrile from *o*-dibromobenzene with copper cyanide in refluxing pyridine. Later, the blue material was identified as copper phthalocyanine. In the following year, the blue impurity in the reaction products was formed during the industrial preparation of phthalimide from phthalic anhydride and ammonia in a glass-lined reaction vessel at the Grangemouth plant of Scottish Dyes Ltd. During the preparation, the glass-lined reaction vessel was cracked. By reason of the reaction carried out, outer steel casing of the reaction vessel, the accident results in the formation of blue impurity. This blue impurity is known to iron phthalocyanine at the present [1–5].

These blue materials determined the molecular structure, which was composed of four iminoisoindoline units with various central metal ions or di-hydrogen by Professor R. P. Linstead at University of London in 1929. Linstead named the by-product *phthalocyanine* as a combination of Greek *naphtha* (rock oil) and *cyanine* (blue) in 1933. The molecular structure of phthalocyanine was confirmed later using X-ray diffraction analysis by Robertson in 1935 [1–5].

Phthalocyanine and metal containing phthalocyanines have been established as blue to green dyestuffs and pigments. Phthalocyanines and metal phthalocyanines are using an important industrial commodity since 1942 [1–5].

Phthalocyanines are an analogous molecular structure as natural colorant of porphyrins. In general, porphyrins consist of four pyrrole units, while phthalocyanines construct four isoindole and nitrogen atoms at *meso* positions. The central cavity of phthalocyanines can place 63 different elemental ions including di-hydrogen (metal-free phthalocyanine). Phthalocyanines containing one or two metal ions are called metal phthalocyanines. In phthalocyanine ring system and part of the atom numbering system, the 2,3,9,10,16,17,23,24 positions are referred to as the peripheral sites and the 1,4,8,11,15,18,22,25 positions as the nonperipheral sites. M can be di-hydrogen or one of the 63 elements of the periodic table (**Figure 1**) [1–5].

As mentioned above, phthalocyanines have been used as green to blue colorants in textile industries because of their thermal, chemical, and photochemical stabilities from their discovery. Over the last decade, phthalocyanines have attractive attention as functional chromophores for various fields such as catalyst, laser light absorbers in data storage systems, electron charge carriers in photocopiers,

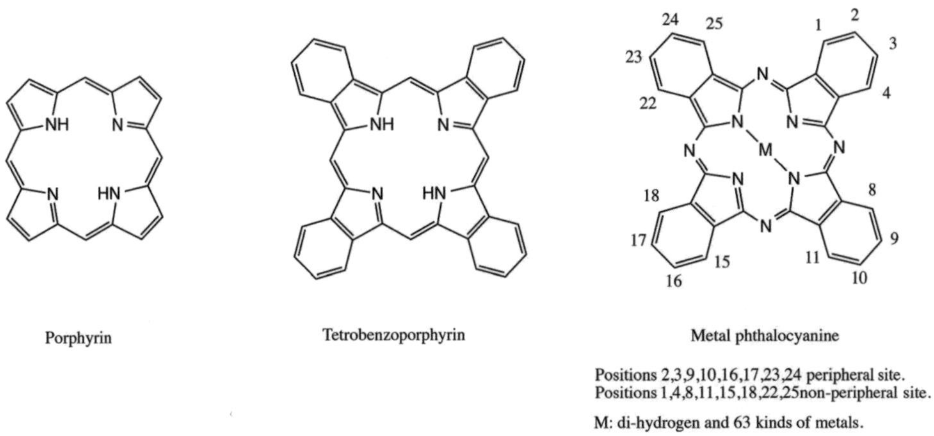

Figure 1.
Molecular structures of porphyrin, porphyrin-related compound, and metal phthalocyanine.

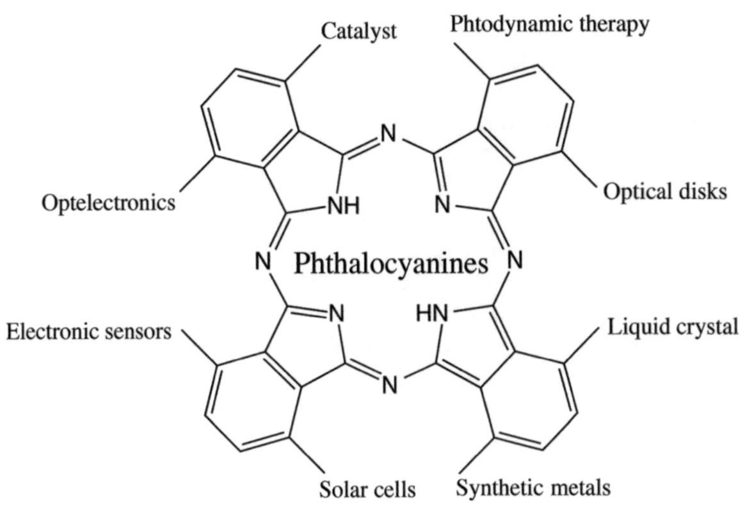

Figure 2.
Typical function of phthalocyanines.

photo-antenna device in photosynthesis, photoconductors in photovoltaic cells, and electrochromic displays, and photosensitizers [6–20] (**Figure 2**).

In order to utilize many applications, the absorption maxima of phthalocyanines are best if moved near infrared region. The strongest absorption of phthalocyanines in visible region called Q band can be attributed to allow from highest occupied molecular orbital (HOMO) to lowest unoccupied molecular orbital (LUMO), which means π–π^* transition. The Q-band of phthalocyanines can be moved by bathochromic effect through extension of the π conjugation system. Especially, phthalocyanines having bathochromic effect are useful for photosensitization purposes.

Photosensitization properties of phthalocyanines are utilized for both photodynamic therapy of cancer (PDT) and dye-sensitized solar cells (DSSCs) [21].

Particularly, phthalocyanines are known to have the potentials to utilize as second-generation photosensitizers for PDT because they have long life time triplet state and show strong absorption of the far-red light between 600 and 850 nm of which a greater penetration of tissue and satisfactory photosensitization of singlet oxygen take place [21–23].

No-substituted phthalocyanines are insoluble or lower solubility in common organic solvents. The weak points of phthalocyanines have been improved to introduce substituents onto the ring system. Alkyl-substituted phthalocyanines become soluble in organic solvents and they have a lipophilic property. The lipophilic phthalocyanines have a high tumor affinity [24]. Hydrophilic-substituted phthalocyanines show solubility in aqueous media. Phthalocyanines containing pyridine rings in place of one or more of the benzenoid rings expected amphiphilic properties [17].

In the second place, phthalocyanines have attractive attention for the conversion of solar to electricity, because dyes come into general used for DSSCs absorb only weakly in solar spectrum. Phthalocyanines for DSSCs are required to possess strong absorption of visible light in the far-red or near infrared region. Then, phthalocyanines have high conversion capability of solar energy to electricity in comparison to common sensitized dyes [25].

In this chapter, synthesis and cyclic voltammetry of soluble phthalocyanines and their homologs compounds, subphthalocyanines were described in order to utilize photosensitizers for PDT and DSSCs [10, 21].

2. Phthalocyanines

2.1 Synthesized peripheral-substituted phthalocyanines

Synthesized phthalocyanines were the followings: phthalocyanine-4,4′,4″4‴-tetrasulfonic acids having sulfonic groups, phthalocyanine-2,3,9,10,16,17,23,24-octacarboxylic acids having carboxylic groups, 2,3,9,10,16,17,23,24-octakis (hexoxymethyl)phthalocyanines and anthraquinocyanines, which has four 9,10-anthraquinone units in the phthalocyanine molecule [26, 27].

Phthalocyanine-4,4′,4″,4‴-tetrasulfonic acids were synthesized from 4-sulfophthalic acid, a metal halide, urea and 1,8-diazabicyclo[5.4.0]undec-7-ene (DBU) as a catalyst [26] (**Figure 3**).

Phthalocyanine-2,3,9,10,16,17,23,24-octacarboxylic acids were synthesized from benzene-1,2,4,5-tetracarboxylic dianhydride (pyromellitic dianhydride), a metal halide and urea under the reaction conditions used for the monomer preparation [26] (**Figure 4**).

The 2,3,9,10,16,17,23,24-octakis(hexoxymethyl)phthalocyanines were synthesized from 1,2-dicyano-4,5-bis(hexoxymethyl)benzene, which was prepared from

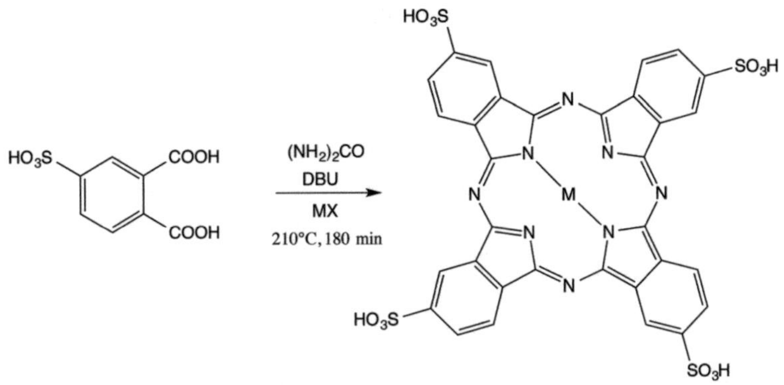

Figure 3.
Synthetic pathway of phthalocyanine-4,4′,4 4 -tetrasulfonic acids.

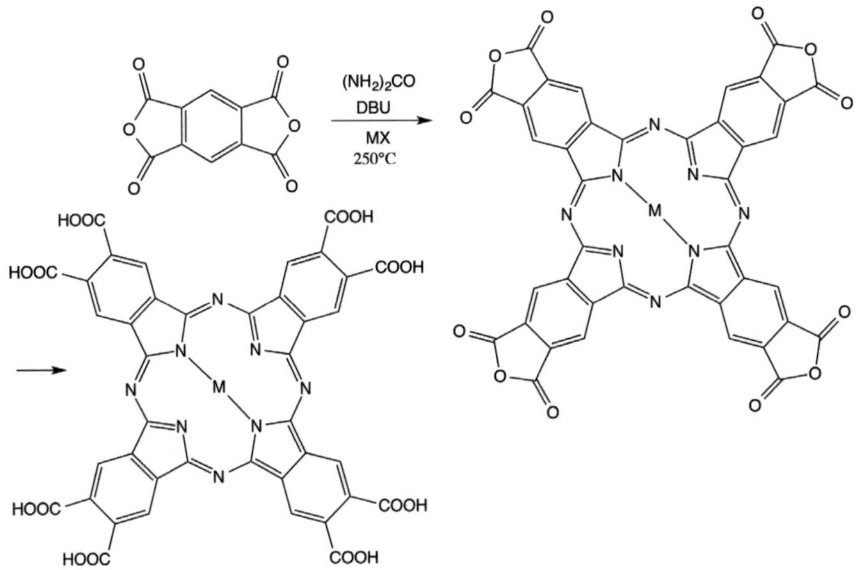

Figure 4.
Synthetic pathway of phthalocyanine-2,3,9,10,16,17,23,24-octacarboxylic acids.

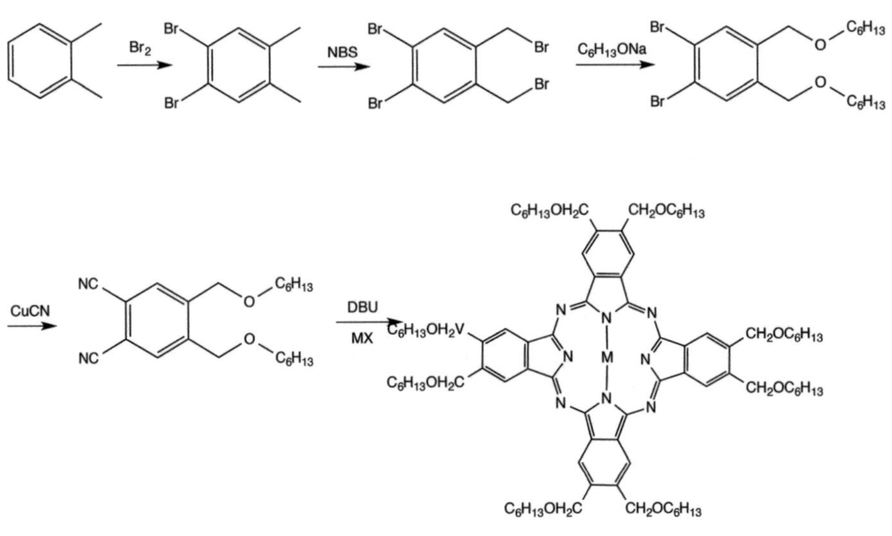

Figure 5.
Synthetic pathway of octakis(hexoxymetyl)phthalocyanines.

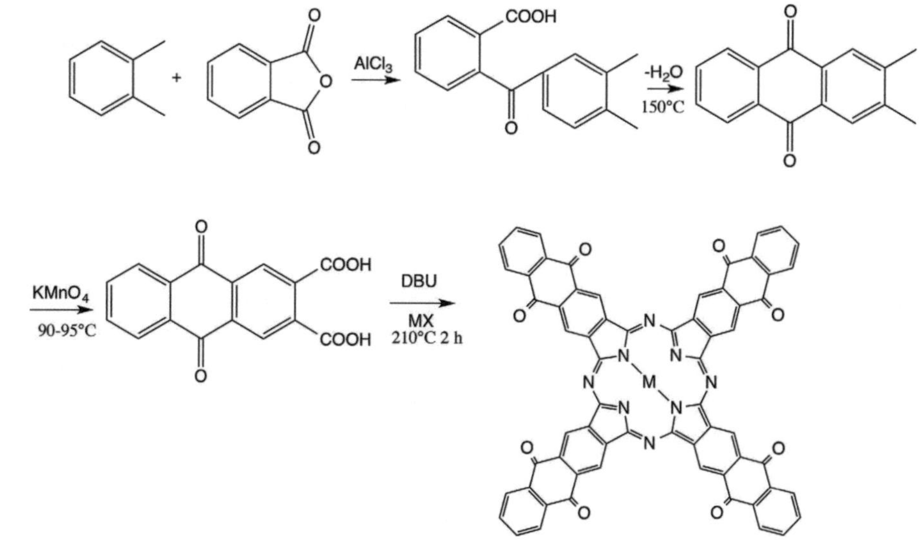

Figure 6.
Synthetic pathway of anthraquinone cyanines.

o-xylene *via* 1,2-dibromo-4,5-dimethylbenzene, 1,2-dibrom-4,5-bis(bromomethyl) benzene, and 1,2-dibromo-4,5-bis(hexoxymethyl)benzene [10] (**Figure 5**).

Anthraquinocyanines were synthesized from 9,10-anthraquinone-2,3-dicarboxylic acid, which was prepared from phthalic anhydride *via o*-(3,4-dimethylbenzoil)benzoic acid and 2,3-dimethyl-9,10-anthraquinone [10] (**Figure 6**).

These phthalocyanines have been measured by cyclic voltammograms (CVs) and chronocoulometric analysis in order to estimate their electron transfer properties and corresponding mechanism.

2.2 Synthesized nonperipheral-substituted phthalocyanines

The author also prepared nonperipheral-substituted phthalocyanine, alkylbenzo-pyridoporphyrazines, which is synthesized by reaction of 3,6-didecylphthalonitrile

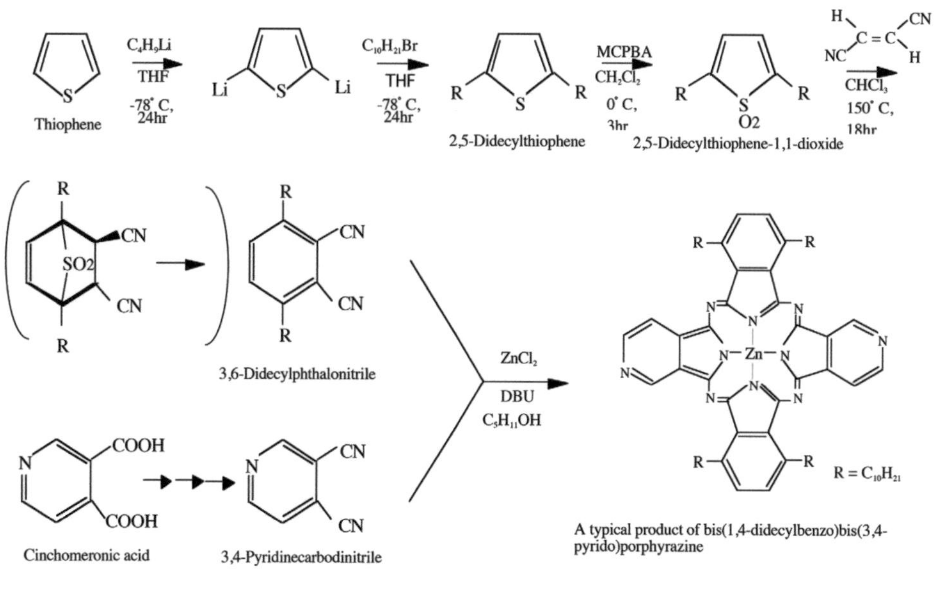

Figure 7.
Synthetic pathway of bis(1,4-didecylbenzo)-bis(3,4-pyrido)porphyrazine.

and 3,4-dicyanopyridine or 2,3-dicyanopyridine in mole ratio of 4:0, 3:1. 1:1, 1:3, and 0:4, respectively. The cross cyclotetramerization product synthesized in mole ratio of 1:1 has been separated with particular attention to give the isolation of regioisomers [17]. Intermediately, 3,6-didecylphthalonitrile was synthesized from thiophene *via* 2,5-didecylthiophene and 2,5-didecylthiophene-1,1-dioxide in accordance with our previous reports [17, 28]. The other intermediates, 3,4-dicyanopyridine and 2,3-dicyanoepyridine were prepared from cinchomeronic acid and quinolinic acid, respectively [17, 27]. The 1:1 mole ratio cross cyclotetramerization products, bis (1,4-didecylbenzo)-bis(3,4-pyrido)porphyrazine and bis(1,4-didecylbenzo)-bis (2,3-pyrido)porphyrazine, were reacted with quaternizing agents such as monochloroacetic acid, diethyl sulfate, and dimethyl sulfate in *N,N*,-dimethyl-formamide as a solvent at 140°C. After quaternation, all compounds gave the water solubility, and got amphiphilic property (**Figures 7–9**).

Nonperipheral arylsulfanyl-substituted phthalocyanines were synthesized in three steps *via* phthalonitrile-3,6-ditriflate and 3,6-bis(arylsulfanyl)phthalonitrile [19, 20, 29–31]. Intermediately, 3,6-bis(arylsulfanyl)phthalonitrile was synthesized from 2,3-dicyanohydroquinone and trifluoromethanesulfonic anhydride for 24 h. Nonperipheral arylsulfanyl phthalocyanines were synthesized from corresponding

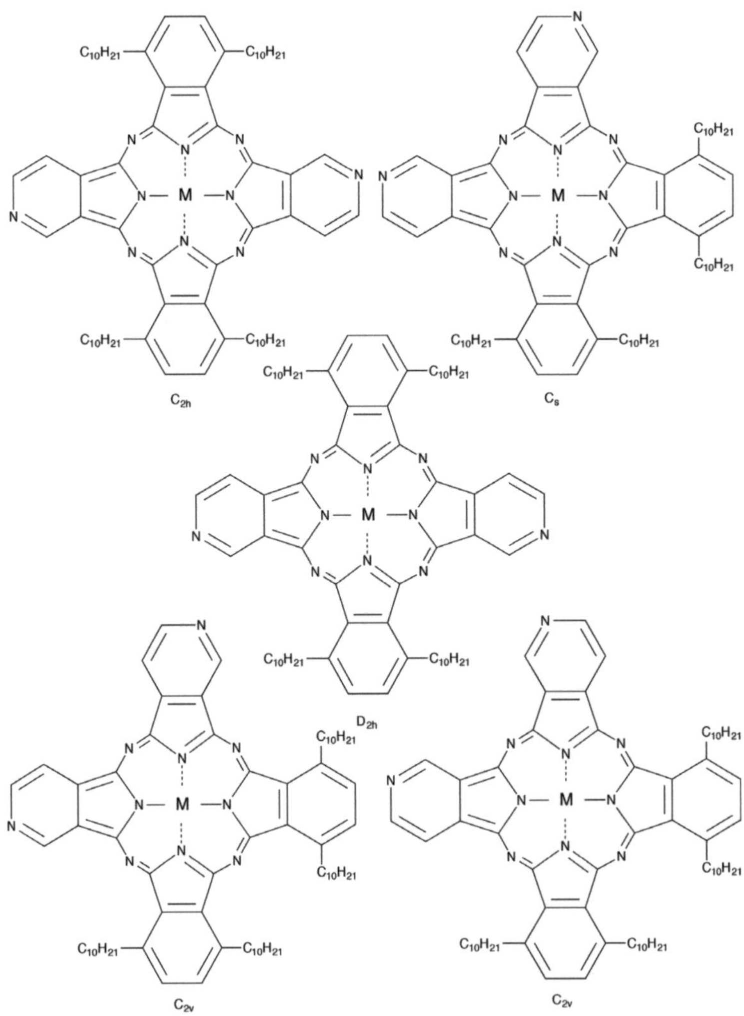

Figure 8.
Regioisomers of metal bis(1,4-didecylbenzo)bis(3,4-pyrido)porphyrazine.

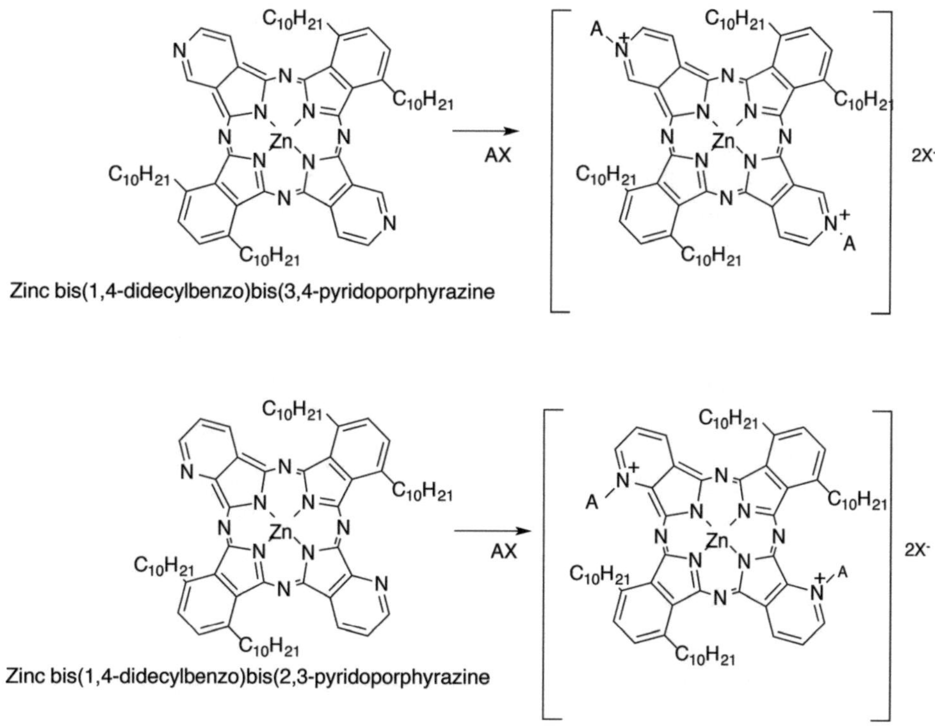

Figure 9.
Quaternation of zinc bis(1,4-didecylbenzo)bis(3,4-pyrido)porphyrazine and zinc bis(1,4-didecylbenzo)bis(2,3-pyrido)porphyrazine; Reagents and conditions: (i) anhydrous $ZnCl_2$, DBU, $C_5H_{11}OH$, 4 h; (ii) AX: monochloroacetic acid, diethyl sulfate or dimethyl sulfate), DMF, 140°C, 2 h.

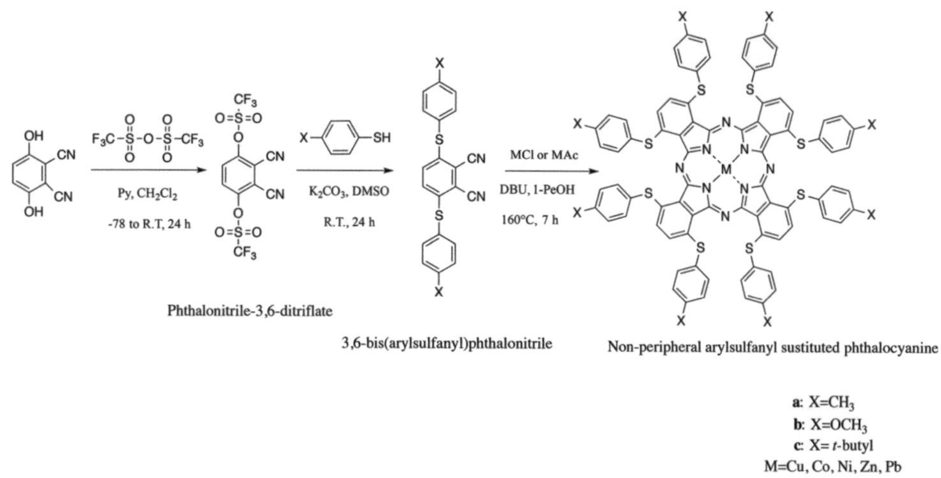

Figure 10.
Synthetic pathway of nonperipheral arylsulfanyl-substituted phthalocyanines.

3,6-bis(arylsulfanyl)phthalonitriles and metal salt in the presence of DBU as a catalyst in 1-pentanol [19, 30, 31] (**Figure 10**).

2.3 Synthesized subphthalocyanines

Subphthalocyanne is the lowest homologous compound of phthalocyanine, which consists of three isoindole units and central boron. Subphthalocyanines have previously been used as reagents for ring enlargement reactions leading to asymmetric phthalocyanines [32] (**Figure 11**).

Subphthalocyanine and four derivatives were synthesized from 1,2-dicyanobenzene or corresponding 1,2-dicyanobenzene derivatives with boron trichloride in 1-chloronaphthalene under argon atmosphere at −3°C [33] (**Figure 12**).

Nonperipheral arylsulfanyl-substituted subphthalocyanines were also synthesized *via* phthalonitrile-3,6-ditriflate and 3,6-bis(arylsulfanyl)phthalonitrile (**Figure 13**) [34].

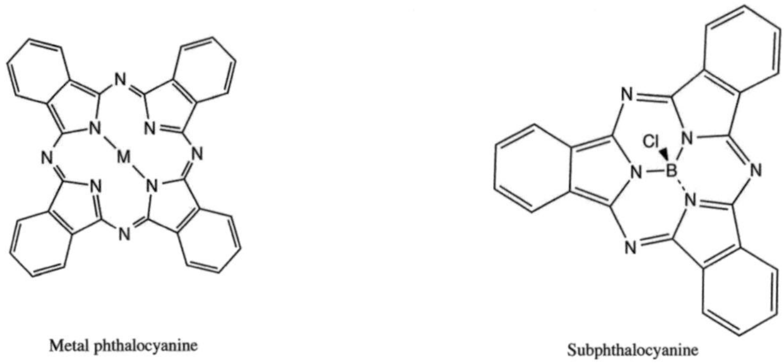

Figure 11.
Molecular structures of metal phthalocyanine and subphthalocyanine.

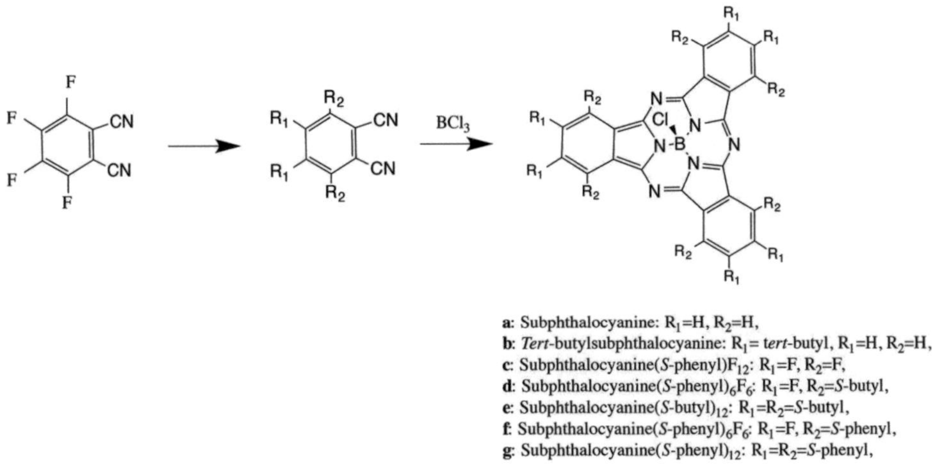

Figure 12.
Synthetic pathway of subphthalocyanine and its six derivatives.

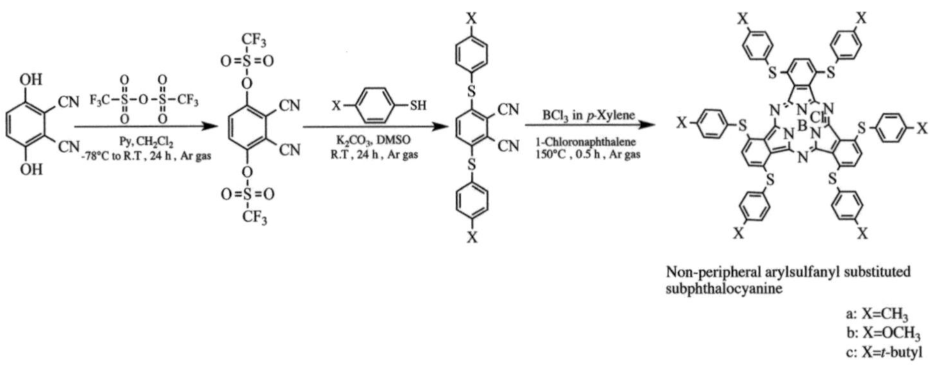

Figure 13.
Synthetic pathway of nonperipheral arylsulfanyl-substituted subphthalocyanines.

DOI: http://dx.doi.org/10.5772/intechopen.81392

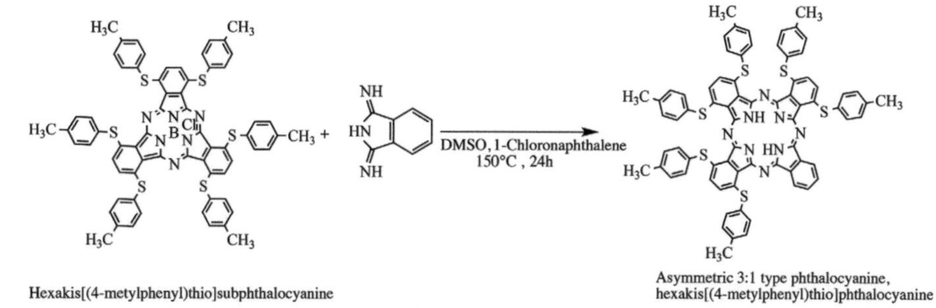

Figure 14.
Ring expansion reaction pathway to prepare asymmetric 3:1 phthalocyanine, hexakis[(4-methylphenyl)thio] phthalocyanine from hexakis[(4-methylphenyl)thio]subphthalocyanine.

To prepare asymmetric 3:1 type phthalocyanines, arylsulfanyl-subphthalocyanines and isoindoline were reacted to obtain metal-free corresponding phthalocyanines (**Figure 14**) [34].

3. Electrochemistry

3.1 Phthalocyanine-4,4′,4″4‴-tetrasulfonic acids and phthalocyanine-2,3,9,10,16,17,23,24-octacarboxylic acids, octakis(hexoxymethyl) phthalocyanine and anthraquinocyanine

CV is used in the estimation of electrochemistry. It is the electrochemical equivalent to spectroscopy. It is a useful tool for the characterization of reduction and oxidation systems. It consists of cyclic potential of a stationary electrode immersed in a quiescent solution and measuring the resulting current. The excitation signal is a linear potential scan with a triangular waveform. This triangular potential excitation signal sweeps the potential of the working electrode. The triangle returns at the same speed and permits the display of a compete voltammogram. Therefore, if a molecular is reduced in the forward scan, it will be re-oxidized on the reverse scan.

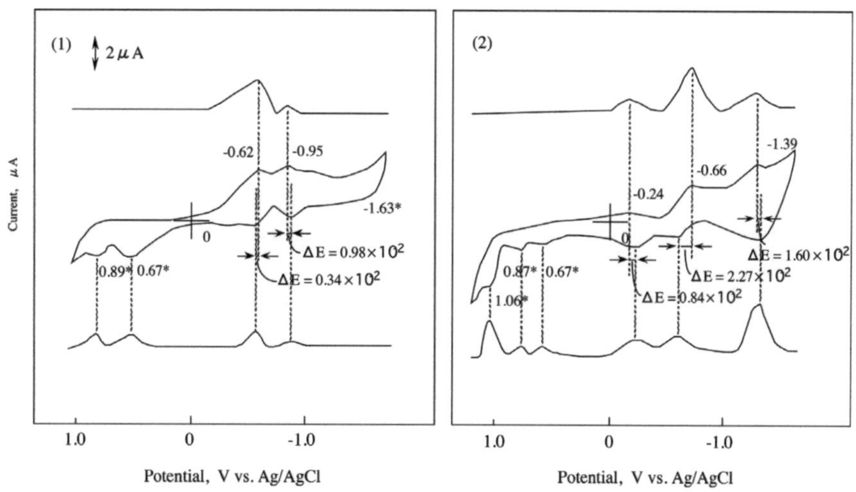

Figure 15.
*Cyclic voltammograms and their first differential curves in in dimethyl sulfoxide with 0.1 mol cm^{-3} tetrabutylammonium perchlorate, scan rate 50 mV s^{-1}. Potentials of the reversible wave are midpoint potential of anodic and cathodic peaks for each couple, $E_{1/2}$; *, irreversible peak; ΔE, the separation between the anodic and cathodic peaks for reversible couple. (1) cobalt phthalocyanine-4,4′,4 ,4 -tetrasulfonic acids, (2) cobalt phthalocyanine-2,3,9,10, 16, 17,23,24-octacarboxylic acids.*

The CV value is the current response, which depends on the applied potential. The current response shows two kinds of peaks such as the upward cathodic and the downward anodic peaks.

Cobalt phthalocyanine-4,4′,4″4‴-tetrasulfonic acids and cobalt phthalocyanine-2,3,9,10,16,17,23,24-octacarboxylic acids are measured electrochemical properties using CV and the first differential curve.

The reported potentials are the midpoint potential of anodic and cathodic peaks for each couple, $E_{1/2}$, and the peak potential for the irreversible step, which have a mark on superscript. The ΔE values are an anodic peak to cathodic peak separation located in the reduction (negative) potential region (**Figure 15**).

The CV of cobalt phthalocyanine-4,4′,4″4‴-tetrasulfonic acids showed two cathodic peaks and four anodic peaks. The peaks are attributed to four reduction stages. The first oxidation potential appeared at 0.67 V versus silver/silver chloride (Ag/AgCl) and the first reversible reduction potential at −0.62 V versus Ag/AgCl. The CV was sorted into five waves. The CVs consist of two reversible reduction couple, one irreversible reduction wave and two irreversible oxidation waves.

The CV of cobalt phthalocyanine-2,3,9,10,16,17,23,24-octacarboxylic acids, three cathodic, and six anodic peaks appeared. The peaks were sorted into three reversible reduction couples at −0.24, −0.66, and −1.39 V versus Ag/AgCl, and three irreversible oxidation waves at 0.67, 0.87, and 1.06 V versus Ag/AgCl. The reduction and oxidation of metal phthalocyanines are due to the interaction between the phthalocyanine macro-ring and the central metal. Sulfonic and carboxylic groups are electron-withdrawing groups, so they are expected to reduce the electro charge in the phthalocyanine macro-ring.

The CV of cobalt 2,3,9,10,16,17,23,24-octakis(hexoxymethyl)phthalocyanine showed four cathodic peaks at 0.16, −0.49, −0.73, and −1.54 V versus Ag/AgCl, and three anodic peaks at 0.73, −0.61, and −1.47 V versus Ag/AgCl. The CV of cobalt 2,3,9,10,16,17,23,24-octakis(hexoxymethyl)phthalocyanine was sorted into two irreversible oxidation waves at 0.16 and 0.73 V versus Ag/AgCl, and two pare of reversible potential. The negative charge is expected to increase on the phthalocyanine macro-ring, since the substituent hexoxymethyl is the electro-donating group.

The CV of cobalt anthraquinocyanine showed a solitary shape in comparison with the other cobalt phthalocyanine-4,4′,4″4‴-tetrasulfonic acids, cobalt phthalocyanine-2,3,9,10,16,17,23,24-octacarboxylic acids, and cobalt 2,3,9,10,16,17,23,24-octakis(hexoxymethyl)phthalocyanine. The shape of CV for cobalt anthraquinocyanine was sorted into three cathodic peaks at 0.19, −0.69, and −0.95 V versus Ag/AgCl, and two anodic peaks at 0.87 and −0.58 V versus Ag/AgCl. Cobalt anthraquinocyanine has almost one pair of reversible potential.

The relationship between the anodic and cathodic peak current ratio of a reversible couple, i_a/i_c and the scan rate, ν, provides a quick test for electrochemical mechanism associated with a preceding or succeeding reversible or irreversible chemical equilibrium. The scan rate varied from 0.05 to 0.3 Vs^{-1} (**Figure 16**).

The ratio of i_a/i_c decreased with an increased ν, all reversible couples of cobalt phthalocyanine-4,4′,4″4‴-tetrasulfonic acids, cobalt phthalocyanine-2,3,9,10,16,17,23,24-octacarboxylic acids, cobalt 2,3,9,10,16,17,23,24-octakis (hexoxymethyl)phthalocyanine, and cobalt anthraquinocyanine. The reversible reduction couples of cobalt phthalocyanine-4,4′,4″4‴-tetrasulfonic acids, cobalt phthalocyanine-2,3,9,10,16,17,23,24-octacarboxylic acids, cobalt 2,3,9,10,16,17,23,24-octakis(hexoxymethyl)phthalocyanine, and cobalt anthraquinocyanine are characterized as a fast reversible electron transfer followed by a reversible chemical reaction. The values of i_a/i_c converge extrapolated to zero of ν, the third reduction potential of cobalt phthalocyanine-2,3,9,10,16,17,23,24-octacarboxylic acids.

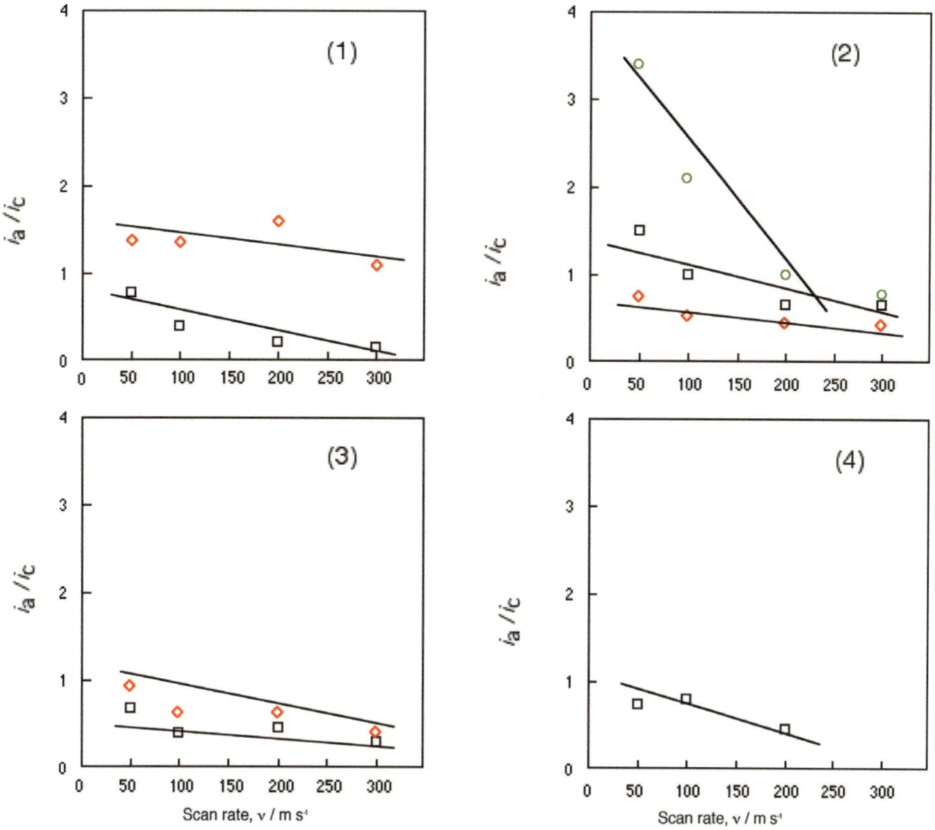

Figure 16.
Change in the anodic to cathodic current ratio with scan rate ν. (1) cobalt phthalocyanine-4,4′,4 ,4 -tetrasulfonic acids; (2) cobalt phthalocyanine-2,3,9,10,16,17,23,24-octacarboxylic acids; (3) cobalt octakis (hexoxymethyl)phtalocyanine; (4) cobalt anthraquinocyanine. Square: first redox couple; Diamond: second redox couple; Circle: third redox couple.

The potentials of ΔE are around 100 mV, except for cobalt phthalocyanine-2,3,9,10,16,17,23,24-octacarboxylic acids. Extrapolated to zero of ν, the ΔE values approach close to 60 mV. The electrode process of cobalt phthalocyanine-4,4′,4″4‴-tetrasulfonic acids, cobalt phthalocyanine-2,3,9,10,16,17,23,24-octacarboxylic acids, cobalt 2,3,9,10,16,17,23,24-octakis(hexoxymethyl)phthalocyanine, and cobalt anthraquinocyanine take place almost one-electron transfer. The $E_{1/2}$ is independent of ν and has constant value. These electrode processes are diffusion-controlled complicated electron transfer having some weak absorption with the oxide (**Table 1**).

Chronoamperometry is a current-time response to a potential step excitation signal. A large cathodic current flows immediately when the potential is stepped up from the initial value, after that it slowly attenuates. The reduction step exhibited that same behavior in comparison with both potential steps. The current-time curves are converted into the relation between the current and square root of time $t^{1/2}$ (**Figures 17** and **18**).

The current-time curve for chronoamperometry is expressed by the Cottrell equation (Eq. (1)).

$$i = \frac{nFCD^{1/2}}{\pi^{1/2}t^{1/2}} = Kt^{1/2} \tag{1}$$

where i is the current (A), n is the number of electrons transferred per ion or molecule (mol^{-1}), F is Faraday's constant (96,485 C mol^{-1}), A is the electrode

Compounds	Potential / V *vs*. Ag/AgCl						
	Reduction				Oxidation		
Cobalt phthalocyanine-4,4',4",4"'-tetrasulfonic acids	-1.63*	-0.95	-0.62		0.67*	0.89*	
ΔE**	0.98	0.34					
Cobalt phthalocyanine-2,3,9,19,16,17,23,24-octacarboxylic acids	-1.39		-0.66	-0.24	0.67*	0.89*	1.06*
ΔE**	1.6		2.27	0.84			
Cobalt octakis(hexoxymethel)phthalocyanine	-1.50	-1.25*	-0.67	-0.49*	0.16*	0.73*	
ΔE**	0.88		1.27				
Cobalt anthaquinocyanine	-0.95*	-0.63			0.19*	0.87*	
ΔE**		1.07					

Potentials of reversible wave are midpoint potential of anodic and cathodic peaks for each couple, $E_{1/2}$.

* Irreversible peak.

** The anodic peak to cathodic peal separation of reversible couple.

Table 1.
Reduction and oxidation potentials.

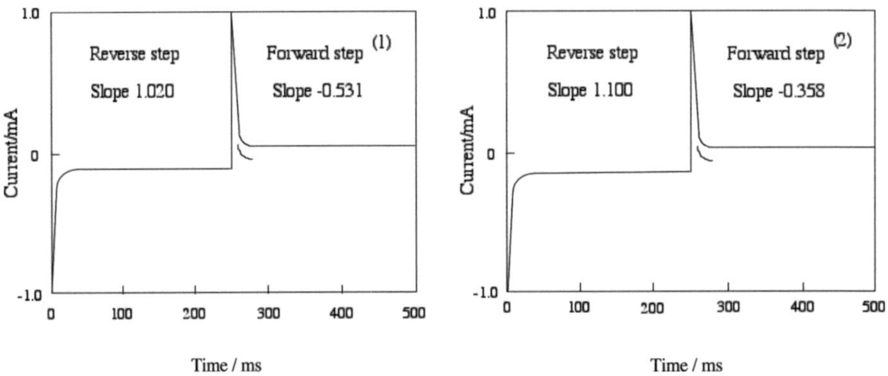

Figure 17.
Chronoamperometry and the slope calculated from the Cottrell plot. Potential step; −1.2 to 1.6 V versus Ag/AgCl, time interval: 250 ms. (1) Cobalt phthalocyanine-4,4′,4 ,4 -tetrasulfonic acids. (2) Cobalt phthalocyanine-2,3,9,10,16,17,23,24-octacarboxylic acids.

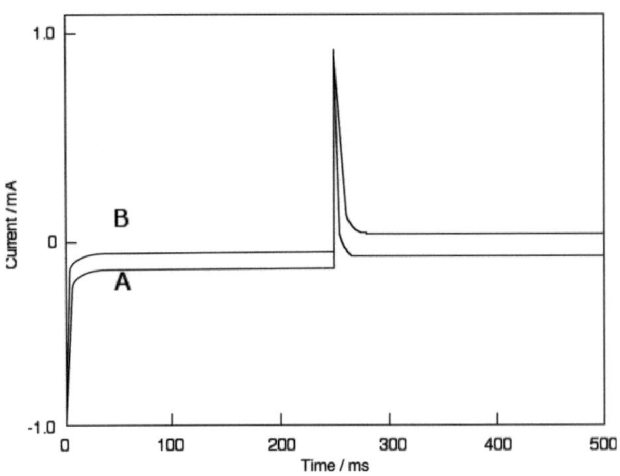

Figure 18.
Chronoamperometry of cobalt oktakis(hxoxymethyl)phthalocyanine. A: pulse step −1.2 to 0 V versus Ag/AgCl, step width 250 ms. B: pulse step −1.2 to 1.6 V versus Ag/AgCl, step width 250 ms.

area ($2.0 \cdot 10^{-2}$ cm^2), C is the concentration (mol cm^{-3}), D is the diffusion constant (cm s^{-1}), and t is time (s). A plot of the current, i versus square root of time, $t^{1/2}$ gives a straight line. The slop means the diffusion constant in forward and reverse steps.

Electron processes in the systems are diffusion-controlled electron transfers mentioned above. Relationships between i and $t^{1/2}$ are considered to be a finite diffusion for cobalt phthalocyanine-4,4′,4″4‴-tetrasulfonic acids and cobalt phthalocyanine-2,3,9,10,16,17,23,24-octacarboxylic acids.

The current of the Cottrell plots is a measure of the rate for electrolysis at the electrode surface. Electrolysis is controlled with a mass transfer by diffusion on the electrode. The diffusion constant implies the rate of electrolysis. The slop means the diffusion constant in each step. The forward step indicates the reduction and the reverse step is oxidation (**Table 2**).

The chronocoulometry was taken by one treatment of chronoamperometry. The current response was integrated to give a response to the charge. The charge-time curve of the forward step for chronocoulometry is the integral of Eq. (1); this is called the Anson equation (Eq. (2)).

$$Q = \frac{2nFACD^{1/2}t^{1/2}}{\pi^{1/2}} = 2Kt^{1/2} \tag{2}$$

The reverse step is following equation (Eq. (3)):

$$Qr = \frac{2nFACD^{1/2}}{\pi^{1/2}}\left\{\tau^{1/2} + (1-r)^{1/2} - t^{1/2}\right\} \tag{3}$$

where τ is time of reverse potential step.

The initial potential was −1.20 V versus Ag/AgCl and the step width was 250 ms. The step potential was 1.60 V versus Ag/AgCl (**Figure 19**).

For cobalt phthalocyanine-4,4′,4″4‴-tetrasulfonic acids, cobalt phthalocyanine-2,3,9,10,16,17,23,24-octacarboxylic acids, cobalt 2,3,9,10,16,17,23,24-octakis (hexoxymethyl)phthalocyanine, and cobalt anthraquinocyanine, the extent of diffusion control increases systematically as the standard potential becomes positive. In the forward step, the electron change reached at about 30 μC. Then, the electron change was decreased 15 μC, except for cobalt phthalocyanine-2,3,9,10,16,17,23,24-octacarboxylic acids. The reverse step was attenuated to 0 μC with 70 ms in the chronocoulometry of the reduction side from −1.20 to 0.00 V versus Ag/AgCl potential, except for cobalt phthalocyanine-2,3,9,10,16,17,23,24-octacarboxylic acids. The chronocoulometry had a linear forward step and a flat reverse step indicating no Faradic activity for all compounds in the oxidation side from the 0.00 to 1.60 V versus Ag/AgCl step.

Compound	Forward step / mV		Reverse step /mV	
	Slop	Intercept	Slop	Intercept
Cobalt phthalocyanine-4,4',4",4'"-tetrasulfonic acids	-0.531	-0.0631	1.020	0.00596
Cobalt phthalocyanine-2,3,9,10,16,17,23,24-octacarboxylic acids	-0.358	-0.1060	1.100	-0.00274
Cobalt octakis(hexoxymethyl)phthalocyanine	-0.525	-0.0360	0.670	-0.000172
Coblt anthraquinocyanine	-0.322	-0.0632	0.698	0.00547

Table 2.
The slop and intercept of the Cottrell plot.

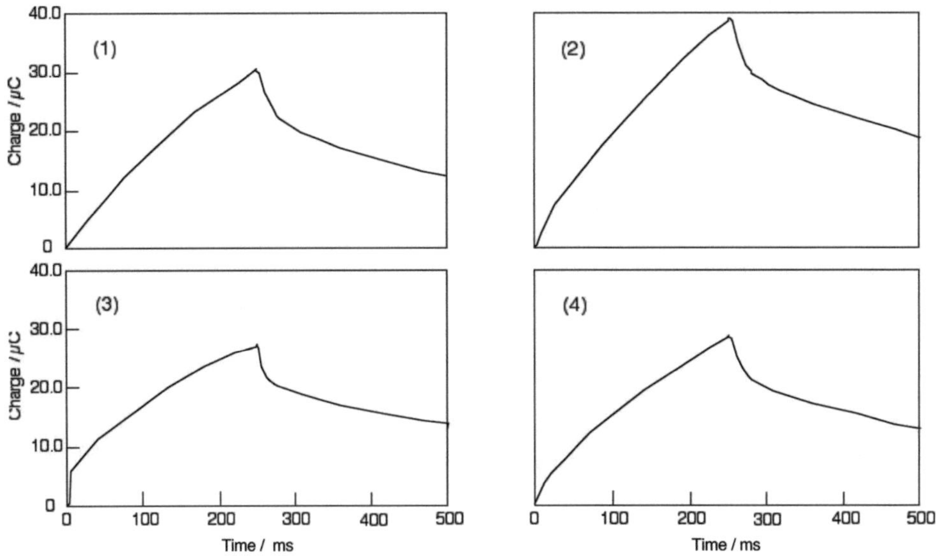

Figure 19.
Chronocoulometry of (1) cobalt phthalocyanine-4,4′,4 ,4 -tetrasulfonic acids, (2) cobalt phthalocyanine-2,3,9,10,16,17,23,24-octacarboxylic acids, (3) cobalt octakis(hexoxymethyl)phthalocyanine, (4) cobalt anthraquinocyanine. Potential step: −1.2 to 1.6 V versus Ag/AgCl, step width 250 ms.

The Anson plots are converted from the charge-time curve of chronocoulometry into the relation between charge and $t^{1/2}$ (**Figure 20**).

The Anson plot is a straight line with an intercept. Chronocoulometry is useful to study absorption on an electrode surface. When absorbed species exist on an electrode surface, it is electrolyzed immediately, whereas solution species must diffuse the electrode in order to react. The total charge Q_{total} is measured in a potential step experiment.

$$Q_{total} = \frac{2nFACD^{1/2}}{\pi^{1/2}} + Q_{dl} + Q_{abs} \tag{4}$$

$$Q_{abs} = nFA\Gamma \tag{5}$$

where Q_{dl} is the double layer charge (C), Q_{abs} is the absorbed species charge (C), Γ is the amount absorbed (mol cm^{-3}). Q_{total} is obtained by summing Q, Q_{abs}, and Q_{dl}. As the expression of Q in Eq. (2), a plot of Q versus $t^{1/2}$ is a straight line. The

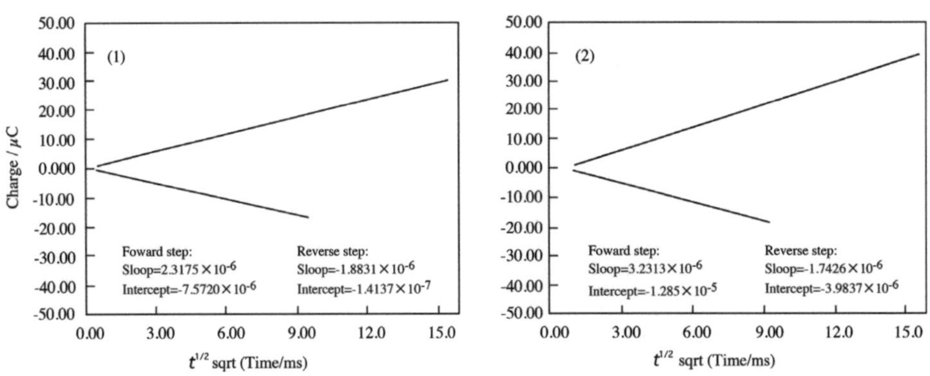

Figure 20.
Anson plots of (1) cobalt phthalocyanine-4,4′,4 ,4 -tetrasulfonic acids and (2) cobalt phthalocyanine-2,3,9,10,16,17,23,24-octacarboxylic acids.

Anson plot should be linear with intercept that is equal to the second and third terms in Eq. (4). If Q_{dl} is known, then, the value Q_{abs} can be calculated for an electrode of the known electrode area. When double step chronocoulometry is used, the difference in the intercepts of forward and reverse steps is Q_{abs}.

Only the value of Q in three terms depends upon the scanning time. The intercept of the Anson plot expresses the sum of Q_{dl} and Q_{abs}. The Q_{abs} can take away Q_{dl}, which is a value of the difference of intercepts between forward and reverse steps, since double step chronocoulometry is used. When no absorption of reactant or product, the intercept of Anson plot for both forward and reverse steps are equal (Q_{dl}). While reactant absorbs but product does not, the intercept of reverse is a measure of Q_{dl} in the presence of absorbed reactant, and the intercept of forward step contains both Q_{dl} and Q_{abs} for absorbed reactant.

The chronocoulometry of cobalt phthalocyanine-4,4′,4″4‴-tetrasulfonic acids, cobalt phthalocyanine-2,3,9,10,16,17,23,24-octacarboxylic acids, cobalt 2,3,9,10,16,17,23,24-octakis(hexoxymethyl)phthalocyanine, and cobalt anthraquinocyanine shows that the reactant is absorbed but not the product. In regard to the absorption using Eqs. (2) and (3), the Q_{abs} calculated to 7.40, 8.92, 2.81, and 7.07 μC for cobalt phthalocyanine-4,4′,4″4‴-tetrasulfonic acids, cobalt phthalocyanine-2,3,9,10,16,17,23,24-octacarboxylic acids, cobalt 2,3,9,10,16,17,23,24-octakis (hexoxymethyl)phthalocyanine, and cobalt anthraquinocyanine, respectively.

The relation between Q_r/Q_f and $t^{1/2}$ can be estimated by the mechanism and rate of the following chemical reaction (**Figure 21**).

The value Q_r/Q_f indicates the base line for the chronocoulometry of the reverse step charge Q_r divided by the final value of forward step Q_f. This relationship can be estimated by the mechanism and rate of following reaction. The following chemical reaction obeyed first-order kinetics, which found the calculation to be 0.20, 0.26, 0.30, and 0.30 s^{-1} for cobalt phthalocyanine-4,4′,4″4‴-tetrasulfonic acids, cobalt phthalocyanine-2,3,9,10,16,17,23,24-octacarboxylic acids, cobalt 2,3,9,10,16,17,23,24-octakis(hexoxymethyl)phthalocyanine, and cobalt anthraquinocyanine [10, 26, 27].

The oxidation of metal phthalocyanines having transition metal are electrochemically irreversible and electrons are added to the orbital of phthalocyanine ring or the central metal depending on the redox potential for reduction process.

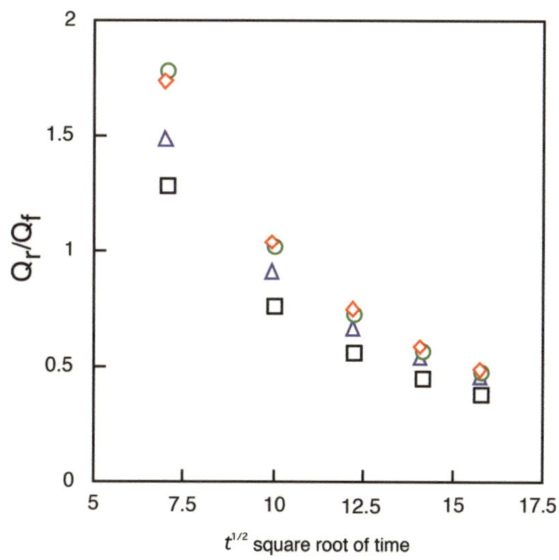

Figure 21.
Variation of Q_r/Q_f with square root of time, $t^{1/2}$ for cobalt anthraquinocyanine.

3.2 Nonperipheral alkylbenzopyridoporphyrazines and nonperipheral arylsulfanyl substituted phthalocyanines

Nonperipheral alkylbenzoporphyrazines synthesized by reaction of 3,6-didecylphthalonitrile and 3,4-dicyanopyridine or 2,3-dicyanopyridine in mole ratio of 4:0, 3:1. 1:1, 1:3, 0:4, respectively. The 1:1 mole ratio cross cyclotetramerization product has been separated with particular attention given to the isolation of regioisomers [17]. At the first time, 3,6-didecylphthalonitrile and 3,4-dicyanopyridine have been reacted together in ratio of 1:1 product, zinc bis (1,4-didecylbenzo)-bis(3,4-pyrido)porphyrazine. The zinc bis(1,4-didecylbenzo)-bis (3,4-pyrido)porphyrazine has two nonperipheral-substituted benzenoido and pyridinoido rings, which are in different locations. The CV can be used to make an estimation of the electrochemical difference for regioisomers (**Table 3**).

Before separation of regioisomers, the reduction and oxidation potentials of zinc bis (1,4-didecylbenzo)-bis(3,4-pyrido)porphyrazine are sorted into six irreversible peaks.

After separation of regioisomers, fractions 1–3 have one pair of reversible oxidation peak and four irreversible waves. Fraction 4 has one pair of reversible and three irreversible waves.

The porphyrazine ring is influenced by the π-electrons about the closed system. Although the π-electron system of zinc bis(1,4-didecylbenzo)-bis(3,4-pyrido) porphyrazine and fractions 1–4 consists of one porphyrazine, two pyridinoid, and two didecyl-substituted benzenoid rings; the location of these rings except for porphyrazine are different from each regioisomer.

Substituents and pyridonoid rings influenced the π-electron environment in zinc bis(1,4-didecylbenzo)-bis(3,4-pyrido)porphyrazine and fractions 1–4. The effect of pyridinoid rings gave rise to the change of the electron density of the metal phthalocyanines. The difference of reduction and oxidation peaks between fractions 1 and 4 is attributed to the effect of variation of the interaction between the central metal and the alkylbenzoporphyrazine. Following this, the difference in CVs between zinc bis(1,4-didecylbenzo)-bis(3,4-pyrido)porphyrazine and fractions 1–4 is also the effect of interaction, since zinc bis(1,4-didecylbenzo)-bis(3,4-pyrido)porphyrazine

Materials	Potential / V *vs.* Ag/AgCl					
	Reduction				Oxidation	
Zinc bis(1,4-didecylbenzo)-bis(3,4-pyrido)porphyrazine	-0.97*	-0.71*	-0.45*	-0.15*	0.37*	0.93*
Fraction 1 D_{2h}	-1.00*	-0.58*	-0.24*		0.44	0.93*
ΔE**					0.17	
Fraction 2 C_{2h}	-1.05*	-0.60*	-0.19*		0.37	0.90*
ΔE**					0.10	
Fraction 3 C_s	-0.96*	-0.65*	-0.22*		0.37	0.89*
ΔE**					0.13	
Fraction 4 C_{2v}	-0.87*	-0.63*	-0.21*		0.34	
ΔE**					0.01	

Zinc bis(1,4-didecylbenzo)bis(3,4-pyrido)porphyrazine was before separation of its regioisomers.
Potentials of reversible wave are midpoint potential of anodic and cathodic peaks for each couple, $E_{1/2}$.
* Irreversible peak.
** The anodic peak to cathodic peak separation for reversible.

Table 3.
Reduction and oxidation potential of zinc bis(1,4-didecylbenzo)bis(3,4-pyrido)porphyrazine and its regioisomers (fractions 1–4).

is a mixture of its regioisomers. The regioisomers of zinc bis(1,4-didecylbenzo)-bis(3,4-pyrido)porphyrazine decided the symmetry of molecular structure as D_{2h}, C_{2h}, C_s and C_{2v} for fractions 1, 2, 3 and 4, respectively. However, two types of C_{2v} isomers of zinc bis(1,4-didecylbenzo)-bis(3,4-pyrido)porphyrazine cannot be isolated.

The ΔE values are around 100 mV and the reduction and oxidation processes are the same for regioisomers, except for fraction 4. The electron process of regioisomers between fractions 1 and 3 involves approximately one electron transfer. The ΔE values of fraction 4 show different behavior in comparison to the others. The different behavior for fraction 4 is attributable to the mixture of two types of C_{2v} regioisomers. The reduction and oxidation potentials of fraction 4 are based on the interaction between two types of C_{2v} regioisomers. No observation on the reversible couple in zinc bis(1,4-didecylbenzo)-bis(3,4-pyrido)porphyrazine resulted in interaction between regioisomers.

Zinc bis(1,4-didecylbenzo)-bis(3,4-pyrido)porphyrazine and zinc bis(1,4-didecylbenzo)-bis(2,3-pyrido)porphyrazine were reacted with quaternizing agents such as monochloroacetic acid, diethyl sulfate, and dimethyl sulfate in *N*,*N*,-dimethylformamide. These compounds were not separated of their regioisomers [10, 17–19].

The shapes of CVs were changed between before and after the quaternation. The reduction and oxidation potentials were shown two anodic and four cathodic peaks for zinc bis(1,4-didecylbenzo)-bis(3,4-pyrido)porphyrazine, and three anodic and four cathodic waves for zinc bis(1,4-didecylbenzo)-bis(2,3-pyrido)porphyrazine (**Figure 22** and **Table 4**).

The redox potential of regioisomers of zinc bis(1,4-didecylbenzo)-bis(3,4-pyrido)porphyrazine were varied (**Table 5**).

After quaternation of regioisomers, the shapes of CVs appeared clearly. The electron transfer ability of regioisomers has been increased remarkably by the acquisition of cation groups. The CV showed two anodic and two cathodic

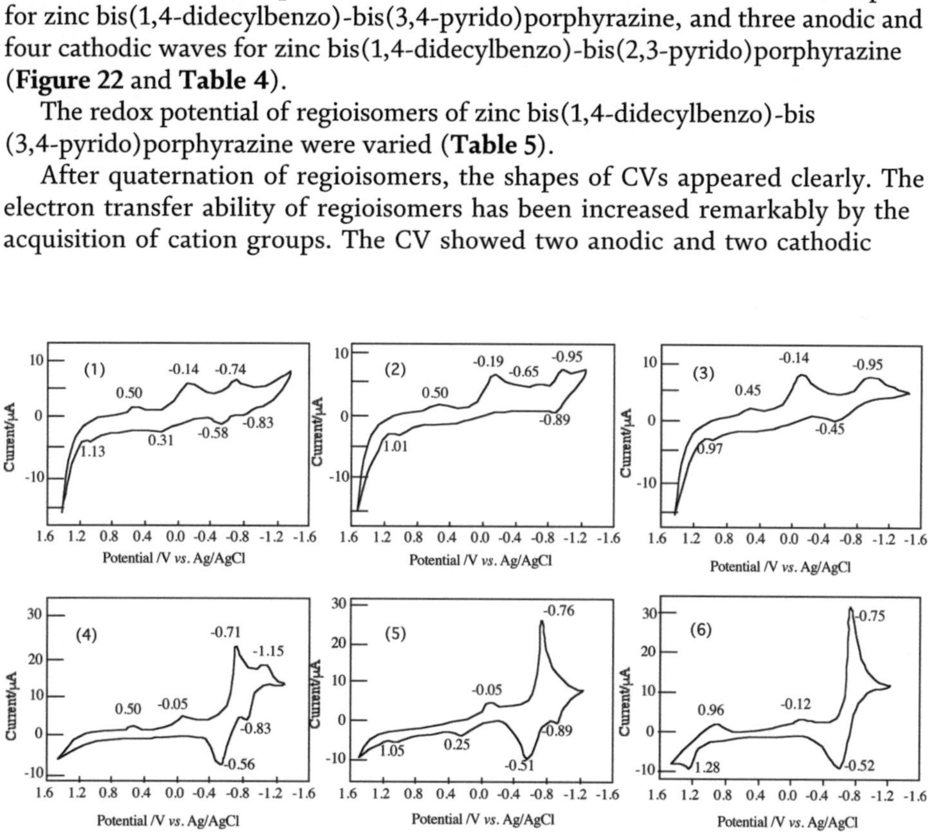

Figure 22.
Potentials of amphiphilic alkylbenzopyridoporphyrazines in N,N-dimethylformamide with tetrabutylammonium perchlorate. (1) bis(1,4-didecylbenzo)-bis(3,4-pyrido)porphyrazine react with dimethyl sulfate, (2) bis(1,4-didecylbenzo)-bis(3,4-pyrido)porphyrazine react with diethyl sulfate, (3) bis(1,4-didecylbenzo)-bis(3,4-pyrido)porphyrazine react with monochloroacetic acid, (4) bis(1,4-didecylbenzo)-bis(2,3-pyrido)porphyrazine react with dimethyl sulfate, (5) bis(1,4-didecylbenzo)-bis(2,3-pyrido)porphyrazine react with diethyl sulfate, (6) bis(1,4-didecylbenzo)-bis(2,3-pyrido)porphyrazine react with monochloroacetic acid.

Compound	Potential / V *vs*. Ag/AgCl						
	Reduction				Oxidation		
Zinc bis(1,4-didecylbenzo)-bis(pyrido)porphyrazine	-0.97*	-0.71*	-0.45*	-0.15*	0.37*	0.93*	
Quaternized with monochloroacetic acid	-0.95*	-0.45*	-0.41*		0.45*	0.97*	
Quaternized with dietyl sulfate	-0.95*	-0.89	-0.65*	-0.19*	0.50*	1.01*	
ΔE**		0.13					
Quaternized with dimethyl sulfate	-0.78	-0.58*	-0.14*		0.31	0.50*	1.13*
ΔE**	0.10						

Potential of reversible wave are midpoint potentials of anodic and cathodic peaks for each couple, $E_{1/2}$.

* Irreversible peak.

** The anodic peak to cathodic peak separation for reversible couple.

Table 4.
Reduction and oxidation potential of zinc bis(1,4-didecylbenzo)-bis (3,4-pyrido)porphyrazine and its quaternized compound with monochloroacetic acid, diethyl sulfate and dimethyl sulfate.

Compounds	Potential / V *vs*. Ag/AgCl				
	Reduction			Oxidation	
Fraction 1 D_{2h}	-0.85*	-0.24		1.02	
ΔE**		0.04			
Fraction 2 C_{2h}	-1.00*	-0.63*	-0.24*	0.47*	0.98*
ΔE**					
Fraction3 C_s	-0.79*	-0.55*	-0.15*	1.09*	
ΔE**					
Fraction 4 C_{2v}	-0.73*	-0.57	-0.31*	0.13*	0.47*
ΔE**		0.07			

Potentials of reversible wave are midpoint potential of anodic and cathodic peaks for each couple, $E_{1/2}$.

* Irreversible peak.
** The anodic peak to cathodic peak separation for reversible couple.

Table 5.
potential of separated regioisomers in zinc bis(1,4-didecylbenzo)-bis(3,4-Pyrido)porphyrazine.

peaks, two anodic and three cathodic peaks, two anodic and two cathodic peaks, and one anodic and five cathodic peaks for fractions 1, 2, 3, and 4, respectively.

The CVs of nonperipheral arylsulfanyl substituted metal-free phthalocyanines such as oxtakis[(4-methylphenyl)thio]phthalocyanine, octakis[(4-methoxyphenyl)thio]phthalocyanine, and octakis[(4-*tert*-butylphenyl)thio]phthalocyanine are of similar shape. The CVs of these metal-free phthalocyanines are known to exhibit irreversible oxidation and reduction potentials around 0.6 and −0.7 V versus the standard hydrogen electrode (SHE), respectively. The reduction and oxidation potentials of phthalocyanines appear from the interaction between the phthalocyanine ring and the central metal atom. In these metal-free phthalocyanines, the shapes of oxtakis[(4-methylphenyl)thio]phthalocyanine, octakis[(4-methoxyphenyl)thio]

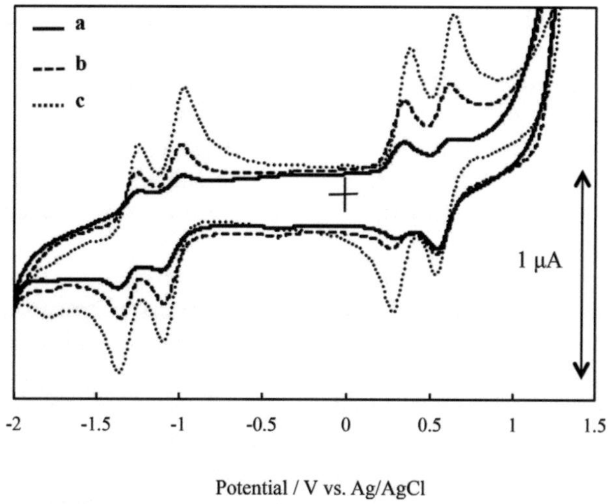

Figure 23.
Cyclic voltammograms of nonperipheral arylsulfanyl-substituted phthalocyanines in odichlorobenzene with 0.1 mol cm^{-3} tetrabutylammonium perchlorate. a: octakis[(4-methylphenyl)thio]phthalocyanine; b: octakis [(4-mthoxylphenyl)thio]phthalocyanine; c: octakis[(4-tert-butyllphenyl)thio]phthalocyanine.

phthalocyanine, and octakis[(4-*tert*-butylphenyl)thio]phthalocyanine showed four cathodic peaks and four anodic peaks. These pairs of peaks arose around −1.3, −1.05, −0.3, and 0.6 V versus Ag/AgCl (**Figure 23**).

The CVs of nonperipheral arylsulfanyl-substituted lead phthalocyanines have similar shapes unconcerned in the presence of terminal groups such as methyl, methoxy, and *tert*-butyl, on the arylsulfanyl substituents. Similar phenomena were observed for other metal phthalocyanines.

Reduction and oxidation potentials of nonperipheral arylsulfanyl-substituted metal-free phthalocyanines and for those having a central metal of cobalt (Co), nickel (Ni), copper (Cu), zinc (Zn), and lead (Pb) were summarized. The CVs of

Central metal	Potential / V vs. Ag/AgCl							
	Reduction				Oxidation			
Octakis[(4-metylphenyl)tio]phthalocyanine-H_2		-1.31	-1.05			0.31	0.58	
Octakis[(4-metylphenyl)tio]phthalocyanine-Co	-1.68	-1.43*		-0.57	0.21		0.55	0.70*
Octakis[(4-metylphenyl)tio]phthalocyanine-Ni	-1.49	-1.32	-1.16	-0.19		0.29	0.57	0.71*
Octakis[(4-metylphenyl)tio]phthalocyanine-Cu		-1.42	-1.13	-0.49*		0.31	0.61	
Octakis[(4-metylphenyl)tio]phthalocyanine-Zn	-1.62	-1.26		-0.70*	0.12		0.53	
Octakis[(4-metylphenyl)tio]phthalocyanine-Pb		-1.29	-1.03	-0.47*	0.19	0.44		
Octakis[(4-methoxyphenyl)tio]phthalocyanine-H_2		-1.31	-1.04			0.31	0.58	
Octakis[(4-methoxyphenyl)tio]phthalocyanine-Co				-0.60	0.17		0.52	0.82
Octakis[(4-methoxyphenyl)tio]phthalocyanine-Ni	-1.49	-1.31	-1.16	-0.18		0.29	0.57	0.68
Octakis[(4-methoxyphenyl)tio]phthalocyanine-Cu	-1.68	-1.43	-1.13			0.28	0.61	
Octakis[(4-methoxyphenyl)tio]phthalocyanine-Zn	-1.61	-1.29			0.06		0.48	
Octakis[(4-methoxyphenyl)tio]phthalocyanine-Pb	-1.54	-1.29	-1.04	-0.47*	0.19	0.44	0.55*	
Octakis[(4-*tert*-butylphenyl)tio]phthalocyanine-H_2		-1.30	-1.03			0.32	0.58	
Octakis[(4-tert-butylphenyl)tio]phthalocyanine-Co	-1.68*	-1.43*		-0.57	0.21		0.53	0.79
Octakis[(4-tert-butylphenyl)tio]phthalocyanine-Ni	-1.53		-0.60	-0.16		0.33	0.59	0.73
Octakis[(4-tert-butylphenyl)tio]phthalocyanine-Cu		-1.43	-1.13			0.30	0.62	
Octakis[(4-tert-butylphenyl)tio]phthalocyanine-Zn		-1.46	-1.19		0.12		0.53	
Octakis[(4-tert-butylphenyl)tio]phthalocyanine-Pb	-1.57*	-1.26	-1.00	-0.35	0.21	0.47	0.69	0.85*

Potentials of reversible wave are midpoint of anodic and cathodic praks for each couple $E_{1/2}$.
*Irreversible peak.

Table 6.
Reduction and oxidation potentials of nonperipheral arylsulfanyl substituted phthalocyanine in o-dichlorobenzene with tetrabutyl ammonium perchlorate.

these compounds exhibited reduction and oxidation potentials, which were in accordance with their central metal (**Table 6**).

The arylsulfanyl substituents in nonperipheral arylsulfanyl-substituted phthalocyanines influence the π electron density in the phthalocyanine ring. The effect of arylsulfanyl groups gives rise to the change of electron density of phthalocyanine ring in the molecule of nonperipheral arylsulfanyl-substituted phthalocyanines. Nonperipheral arylsulfanyl-substituted phthalocyanines exhibit excellent electron transfer properties.

The reduction and oxidation properties of nonperipheral arylsulfanyl-substituted phthalocyanines result from electron transfer from sulfur atoms in the arylsulfanyl substituents at the peripheral positions of phthalocyanine ring to the central metal atom, except in the case of Co as the central metal. In nonperipheral arylsulfanyl-substituted Co phthalocyanines, the irreversible peaks are attributed to the central metal and the reversible waves represent the reduction/oxidation of phthalocyanine ring, including arylsulfanyl groups. It appears that the electron transfer mechanism of nonperipheral arylsulfanyl-substituted phthalocyanines depend on the kind of central metal atom. In particular, the electron transfer mechanism of nonperipheral arylsulfanyl-substituted Co phthalocyanines and nonperipheral arylsulfanyl-substituted Zn phthalocyanines were attributed to reduction and oxidation of their central metal, while those of nonperipheral arylsulfanyl-substituted metal-free phthalocyanines, nonperipheral arylsulfanyl-substituted Cu phthalocyanines, nonperipheral arylsulfanyl-substituted Ni phthalocyanines, and nonperipheral arylsulfanyl-substituted Pb phthalocyanines resulted from HOMO to LUMO electron transitions [20].

3.3 Subphthalocyanines and nonperipheral arylsulfanyl-substituted subphthalocyanines

The CV of subphthalocyanine showed two cathodic peaks at −0.30 and −0.65 V versus Ag/AgCl, and two anodic peaks at 0.82 and −0.62 V versus Ag/AgCl. Subphthalocyanine has two irreversible oxidation and reduction at 0.82 and −0.30 V versus Ag/AgCl, and one pair of reversible reduction potential at −0.64 V versus Ag/AgCl (**Figure 24**).

The reduction and oxidation potentials of subphthalocyanine and its derivatives are summarized (**Table 7**).

In the case of metal phthalocyanines, their substituents influence the π electron environment in the molecule especially the four phenylene rings. The effect of substituents gives rise to the change of electron density of the four phenylene rings in the molecule. Electron transfer properties depend on the kind of the substituents.

While in the case of subphthalocyanine and its derivatives, reduction and oxidation potentials were of various values. However, one irreversible potential certainly appeared around −0.3 V versus Ag/AgCl, of which peaks are attributed to the reduction of the subphthalocyanine ring. The difference of CV between subphthalocyanine derivatives is attributed to the variation of the substituents depends on the subphthalocyanine ring [33].

The CVs of nonperipheral arylsulfanyl-substituted subphthalocyanines show similar shapes (**Figure 25**).

The shapes of hexakis[(4-*tert*-butylphenyl)thio]subphthalocyanine showed cathodic and anodic peaks. The pair of peaks appeared at around −1.69, −1.15, 0.62, and 0.88 V versus Ag/AgCl. Oxidation and reduction potentials of nonperipheral arylsulfanyl-substituted subphthalocyanines were summarized (**Table 8**).

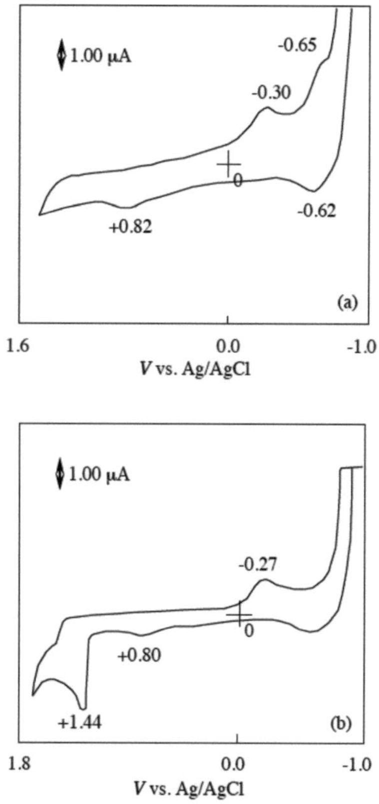

Figure 24.
Cyclic voltammograms of unsubstituted subphthalocyanine and dodecylkis(thiobutyl)subphthalocyanine in acetonitrile with 0.1 mol cm^{-3} tetrabutylammonium perchlorate. (a) unsubstituted SubPC, (b) dodecylkis (thiobutyl)subphthalocyanine.

As mentioned above, in general, metal phthalocyanines having transition metal behave as electrochemically irreversible, and exhibit reduction and oxidation properties resulting from interaction between the phthalocyanine ring and their central metal. The oxidation potential is about 1.0 V versus SHE. The reduction potential occurs between −0.3 and −0.8 V versus SHE. Electrons are added either to the molecular orbital of the phthalocyanine ring or the central metal, depending on the reduction and oxidation potential for reduction process.

As mentioned above, an irreversible reduction appeared around −0.3 V versus Ag/AgCl for subphthalocyanines [33]. The irreversible peaks of subphthalocyanines are attributed to the reduction of the subphthalocyanine ring. The reduction and oxidation potentials of subphthalocyanines result from their substituents.

CVs of nonperipheral arylsulfanyl-substituted subphthalocyanines differ in terms of shape from subphthalocyanine; hexakis[(4-*tert*-butylphenyl)thio] subphthalocyanine has many reduction and oxidation peaks. These phenomena mean that the substituents variedly act in accordance with their electro-donating property for methyl, methoxy, and *tert*-butyl, although they do not demonstrate the effect as chromophores. The substituents of nonperipheral arylsulfanyl-substituted subphthalocyanines are affected more by subphthalocyanines than phthalocyanine ring above mentioned corresponding octakis (arylsulfanyl)phthalocyanines [34]. Nonperipheral arylsulfanyl-substituted subphthalocyanines show many reduction potentials. Nonperipheral arylsulfanyl-substituted subphthalocyanines are acceptable electrons in the

Compound	Potential / V vs. Ag/AgCl			
	Reduction		Oxidation	
Hexakis(thiobutyl)hexafuluoro-subphthalocyanine	0.56*	-0.28*	0.83	
DE^b	0.03			
Hexakis(thiophenyl)hexafuluoro-subphthalocyanine	-0.27*		0.80*	1.44*
Dodecylkis(thiobutyl)hexafuluoro-subphthalocyanine	-0.21*	-0.37*	0.18*	
Dodecylkis(thiophenyl)hexafuluoro-subphthalocyanine	-0.57*	-0.46*	0.18*	
Subphthalocyanine	-0.64	-0.03*	0.82*	
Δ*E***	0.03			

Potentials of reversible wave are midoopoint potential of anodic and cathodic peaks for each couple $E_{1/2}$.

* Irreversible peak.

** The anodic peak to cathodic peak separation for reversible couple.

Table 7.
Reduction and oxidation potential of subphthalocyanines.

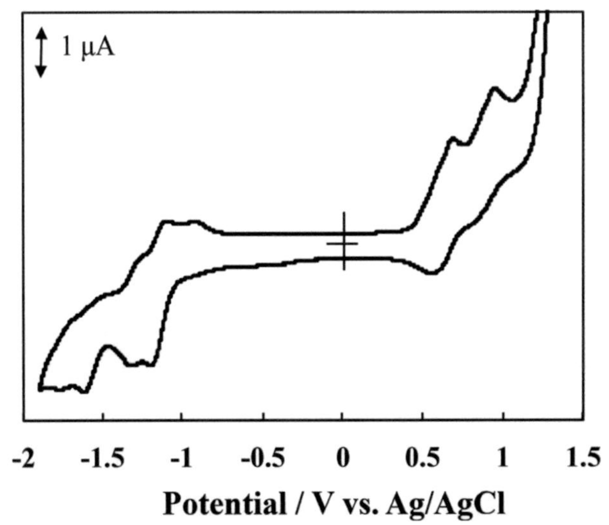

Figure 25.
Cyclic voltammogram of hexakis[(4-tert-butyl)thio]subphthalocyanine in o-dichlorobenzene with 0.1 mol cm^{-3} tetrabutylammonium perchlorate.

Compound	Potential / V vs.Ag/AgCl in dichloromethane Solvent						
	Reduction					Oxidation	
Hexakis[(4-metylphenyl)thio]subphthalocyanine	-1.74	-1.56	-1.28	-1.14	-0.92*	0.62	0.90
Hexakis[(4-*tert*-butylphenyl)thio]subphthalocyanin		-1.69	-1.15	-0.90*		0.62	0.88
Hexakis[(4-metylphenyl)thio]phthalocyanine		-1.25	-0.86	-0.46		0.38	
Subphthalocyanine			-1.31			0.75	
	Potential / V vs.Ag/AgCl in chloroform Solvent						
	Reduction					Oxidation	
Hexakis[(4-metylphenyl)thio]subphthalocyanine			-1.81	-1.43*		0.80*	
Hexakis[(4-tert-butylphenyl)thio]subphthalocyanine			-1.81			0.74*	
Subphthalocyanine			-1.52*	-1.20*		0.81*	

Potentials of reversible wave are midpoint of anodic and cathodic praks for each couple $E_{1/2}$.

*Irreversible peak.

Table 8.
Reduction and oxidation potential of nonperipheral arylsulfanyl-subphthalocyanine and asymmetric 3:1 phthalocyanine, hexakis(arylsulfanyl)phthalocyanine in o-dichlorobenzene or chloroform with 0.1 mol cm^{-3} tetrabutylammonium perchlorate.

subphthalocyanine ring compare with corresponding octakis(arylsulfanyl) phthalocyanines [34].

CV of asymmetric 3:1 type phthalocyanine, hexakis[(4-methylphenyl)thio] phthalocyanine shows peaks at −1.25, −0.86, 00.46, and 0.38 V versus Ag/AgCl. These data are similar to octakis[(4-methylphenyl)thio]phthalocyanine.

4. Conclusion

Electrochemical measurements were performed cobalt phthalocyanine-4,4′,4″ 4‴-tetrasulfonic acids, cobalt phthalocyanine-2,3,9,10,16,17,23,24-octacarboxylic acids, cobalt 2,3,9,10,16,17,23,24-octakis(hexoxymethyl)phthalocyanine, and cobalt anthraquinocyanine in order to examine their electron transfer abilities and electrochemical mechanism.

The CVs of zinc bis(1,4-didecylbenzo)-bis(3,4-pyrido)porphyrazine and its regioisomers showed different oxidation and reduction potentials. After quaternation, the shapes of CVs appeared clearly. It is though that electron transfer ability has been increased remarkably by acquisition of cation groups.

The electron transfer properties of nonperipheral arylsulfanyl-substituted phthalocyanines were shown to be excellent estimated. The effect of arylsulfanyl groups gives rise to be significant.

Electron transition in nonperipheral arylsulfanyl-substituted Co phthalocyanines and nonperipheral arylsulfanyl-substituted Zn phthalocyanines are attributed to reduction and oxidation of their central metal, whereas, in nonperipheral arylsulfanyl-substituted metal-free phthalocyanines, nonperipheral arylsulfanyl-substituted Cu phthalocyanines, nonperipheral arylsulfanyl-substituted Ni phthalocyanines and nonperipheral arylsulfanyl-substituted Pb phthalocyanines, electron transfer results from HOMO to LUMO electron transition.

For subphthalocyanine and its derivatives, reduction and oxidation potentials were various values. The difference of CV between subphthalocyanine derivatives is attributed to the variation of the substituents depends on the subphthalocyanine ring.

Nonperipheral arylsulfanyl-substituted subphthalocyanines show many reduction potentials. Nonperipheral arylsulfanyl-substituted subphthalocyanines are acceptable electrons in the subphthalocyanine ring, meaning that the nonperipheral arylsulfanyl-substituted subphthalocyanines have good electron transfer properties.

Asymmetric 3:1 type phthalocyanine, hexakis[(4-methoxyphenyl)thio]phthalocyanine has similar properties of previous reported phthalocyanines.

These results suggest that phthalocyanines will be appropriate materials for use in the next generation photosensitize for PDT and DSSCs.

Acknowledgements

I thank Emeritus Professor Dr. Michael J. Cook (University of East Anglia, UK), the late Emeritus Professor Dr. Ryo Hirohashi (Chiba University, Japan), Emeritus Professor Dr. Nagao Kobayashi (Tohoku University, Japan), Professor Dr. Victor N. Nemykin (University of Minnesota, USA) and Dr. Toshiyuki Urano (Science and Technology Research center, MitubishiChemical Corporation, Japan) for their interest and helpful advice. I am grateful to my previous graduate students Dr. Eiko Ohno-Okumura (Former: Research Institute of Chemical Science, Technology and Education, Japan. Currently: Tokyo Chemical Industry, Japan), Mr. Taku Kato (Electric Materials Research Laboratories, Nissan Chemical Industry, Japan),

Mr. Masayoshi Kawaguchi, Ms. Tomomi Kawguchi, Mr. Masaki Watanabe (T. S. Insturuments, Japan), Ms. Seiko Urata-Kanazawa (Information and Communication Division, Toppan Printing, Japan), Mr. Hisashi Soga (Riken, Japan), Mr. Naoki Furuya, Mr. Kazuhiro Sugaya, Mr. Makoto Takemoto, Ms. Hitomi Kubo, Mr. Shouta Watabiki, Mr. Yuma Skaguchi and Ms. Yumiko Igarashi. I am pleased to acknowledge the considerable assistance of Mr. Tomoyuki Ogawa (Central Research Center, Nissan Chemical Industry, Japan). The assistance of Dr. Tomoe Komoriya, Associate Professor at Nihon University, and Dr. Satoru Yoshino, Adjunct Professor at Nihon University are greatly appreciated. I am also thankful to have the assistance of my undergraduate students who contributed their experimental skills, sustained effort, and grasp of objectives to accomplishment of experimental program.

Author details

Keiichi Sakamoto[1,2]

1 Department of Sustainable Engineering, College of Industrial Technology, Nihon University, Chiba-ken, Japan

2 Academic Major of Applied Molecular Chemistry, Graduate School of Industrial Technology, Nihon University, Chiba-ken, Japan

*Address all correspondence to: sakamoto.keiichi@nihon-u.ac.jp

References

[1] Hirohashi R, Sakamoto K, Ohno-Okumura E, editors. Phthalocyanines as Functional Dyes. Tokyo: IPC; 2004. pp. 1-22

[2] McKeown NB. Phthalocyanine Materials, Synthesis, Structure and Function. Cambridge: Cambridge University Press; 1998. pp. 1-11

[3] Lezonoff CC, Lever ABP, editors. Phthalocyanines. Vol. 1. Ontario: VCH Publishers; 1989. pp. 5-6

[4] Moser FH, Thomas AL. The Phthalocyanines. Vol. 1. Florida: CRC Press; 1983. pp. 1-3

[5] Moser FH, Thomas AL. Phthalocyanine Compounds. New York: Chapman & Hall; 1963. pp. 1-9

[6] Likyanets EA, Nemykin VN. Journal of Porphyrins and Phthalocyanines. 2010;**14**:1-40

[7] Campidell S, Ballesteros B, Filoramo A, Diaz D, de la Torre G, Torres T, et al. Journal of the American Chemical Society. 2008;**139**:11503-11509

[8] de la Torre G, Vazques P, Torres T. Chemical Reviews. 2004;**104**:3723-3750

[9] Cid JJ, Jang SR, Nazeeruddin MK, Matinez-Ferrero E, Ko J, Graetzel M, et al. Angewandte Chemie, International Edition. 2007;**46**:8358-8362

[10] Sakamoto K, Ohno-Okumura E. Materials. 2009;**2**:1127-1180

[11] Pinzon JR, Plnska-Brzezinska ME, Cadona CM, Athans AJ, Gayathri SS, Guldi DM, et al. Angewandte Chemie, International Edition. 2008;**47**: 4173-4176

[12] Drumus M, Taman H, Gol C, Ahsen V, Nykong T. Dyes and Pigments. 2011; **91**:153-156

[13] Ichikawa M, Kobayashi K, Koyama T, Taniguchi Y. Thin Solid Films. 2007; **515**:3932-3935

[14] Shao J, Thomas A, Han B, Hansen CA. Journal of Porphyrins and Phthalocyanines. 2010;**14**:133-141

[15] Sharma U, Verma PK, Kumar V, Bala M, Singh B. Chemistry–A European Journal. 2011;**17**:5903-5907

[16] Bertagnolli H, Krishnan V, Youssef TE, Hanack M. Journal of Porphyrins and Phthalocyanines. 2011;**15**:598-601

[17] Sakamoto K, Kato T, Cook MJ. Journal of Porphyrins and Phthalocyanines. 2001;**5**:742-750

[18] Sakamoto K, Kato T, Kawaguchi T, Ohno-Okumura E, Urano T, Yamaoka T, et al. Journal of Photochemistry and Photobiology A: Chemistry. 2002;**153**: 245-253

[19] Sakamoto K, Ohno-Okumura E, Kato T, Soga H. Journal of Porphyrins and Phthalocyanines. 2010;**14**:47-54

[20] Sakamoto K, Furuya N, Soga H, Yoshino S. Dyes and Pigments. 2013;**96**: 430-434

[21] Sakamoto K, Yoshino S, Takemoto M, Furuya N. Journal of Porphyrins and Phthalocyanines. 2013;**17**:605-627

[22] Haywood-Small SL, Vernon DI, Griffiths J, Schofield J, Brown SB. Biochemical and Biophysical Research Communications. 2006;**339**:569-576

[23] Komiarov AN, Makarov DA, Yuzhakova OA, Savvina LP, Kuzznetsiva NA, Kaliya OL, et al. Macroheterocycles. 2012;**5**:169-174

[24] Jori G. Photochemistry and Photobiology. 1990;**52**:439-443

[25] Idown M, Loewenstein T, Hastall A, Nykong T, Schlettwein d. Journal of Porphyrins and Phthalocyanines. 2010; **14**:142-149

[26] Sakamoto K, Ohno E. Journal of the Society of Dyers and Colourists. 1996; **112**:368-374

[27] Ohno E, Sakamoto K. The Chemical Society of Japan. 1995;**1995**:730-735

[28] Sakamoto K, Ohno-Okumura E, Kato T, Watanabe M, Cook MJ. Dyes and Pigments. 2008;**78**:213-218

[29] Sakamoto K, Kato T, Ohno-Okumura E, Watanabe M, Cook MJ. Dyes and Pigments. 2005;**64**:63-71. DOI: 10.1016/j.dypig.2007.12.004

[30] Sakamoto K, Watabiki S, Yoshino S, Komoriya T. Journal of Porphyrins and Phthalocyanines. 2017;**21**:658-664

[31] Sakamoto K, Yoshino S, Takemoto M, Sugaya K, Kubo H, Komoriya T, et al. American Journal of Analytical Chemistry. 2014;**5**:1037-1045

[32] Claessens CG, Gonzalez-Rodriguez D, Torre T. Chemical Reviews. 2002; **102**:835-853

[33] Ohno-Okumura E, Sakamoto K, Kato T, Hatano T, Fukui K, Karatsu T, et al. Dyes and Pigments. 2002;**53**:57-65

[34] Sakamoto K, Yoshino S, Takemoto M, Sugaya K, Kubo H, Komoriya T, et al. Journal of Porphyrins and Phthalocyanines. 2015;**19**:688-694

Chapter 4

Electrochemical Preparation and Characterization of Chemically Modified Electrodes

Kenneth L. Brown

Abstract

This chapter will focus on the use of electropolymerization schemes and reactions to modify electrode surfaces with organo- and organometallic compounds such as metal(II)-4, 9, 16, 23-tetraminophthalocyanines and metal complexes of 5-amino-1, 10-phenanthroline. The chemical modification described herein focuses on the formation of polymer films on electrode surfaces. In addition, the characterization of such films using cyclic voltammetry, chronoamperometry, chronocoulometry, and electrochemical impedance spectroscopy will be a key emphasis within this chapter. A brief review of the literature concerning these techniques along with relevant compounds that have been used in studies related to these voltammetric techniques is presented.

Keywords: electropolymerization, cyclic voltammetry, polymer films

1. Introduction

Chemically modified electrodes (CMEs) over the last decade has continued to spark considerable interest in analytical chemistry with respect to electrocatalysis, development of new electrochemical and spectroelectrochemical sensors, and electrochromic displays, just to name a few. Since the novel work by Lane and Hubbard involving chemisorbed metals on platinum electrode surfaces, numerous methods have been developed to immobilize species onto a variety of electrode surfaces. These modification techniques include covalent attachment, spin coating, electropolymerization, and others.

2. Significance of electrode modification

The significance and role of chemically modified electrodes (CMEs) to catalyze or modify the rate of chemical reactions has been extensively surveyed for a number of applications [1–4] . The methodology for modifying an electrode surface has a significant impact on its short-term and long-term use in analytical applications. Many of these applications involve the detection of different types of analytes, including transition metal ions, peroxides, anions, and organic species [5–8]. Understanding the properties of a chemically modified electrode such as mechanisms of charge transfer, electrode stability in different chemical environments such as low and high pH certainly lays a foundation for developing methods based on certain types of reactions (e.g., chemiluminescence).

Immobilized species on electrode surfaces can be composed of several monolayers or several thousand monolayers, as in the case of electropolymerized thin films [1]. The electrode modification procedure is tailored to control charge transfer rates for specific applications such as (1) electrocatalysis, (2) solar energy conversion, (3) improving electrode stability, (4) development of sensors/biosensors, (5) directed and controlled mediated charge transfer, and (6) electrochromism for color displays which are potential dependent [9–13]. Such applications are usually based upon the charge-transfer properties of the compound immobilized on the electrode. Presently, many applications involve the use polymeric multilayered thin films, which may be loaded ionomers, redox polymers, or electronically conducting polymers. Some of the methods for modifying electrodes include chemisorption, covalent binding, vacuum deposition, electroplating, electropolymerization, and others [14–20].

One of the benefits of using electropolymerization reactions to modify electrode surfaces is the ability to change the properties of the polymer film by changing the experimental conditions (e.g., scan rate in cyclic voltammetry); the ability to chemically modify irregular shaped electrodes, and the ability to tailor an electrode surface for specific applications.

Electropolymerized thin films as selective reagents on various surfaces have been at the forefront in designing amperometric sensors. Two significant advantages of using electropolymerized thin films and coatings in electrochemical amperometric sensors and spectroelectrochemical sensors are the ability to (1) better control the chemical and physical nature of the reaction system, and (2) provide a much higher "effective concentration" per unit area for reactions to occur. The challenge in designing polymeric films is tailoring the molecular architecture to control and direct chemical reactivity with selected compounds [17, 18]. One of the most widely used methods in developing polymeric films on various electrode surfaces is electropolymerization via cyclic voltammetry [19, 20]. This technique, under consistent experimental conditions, produces a polymeric film consistent in molecular architecture and reactivity toward specific compounds.

The electrochemical reactivity of specific compounds toward polymeric films is influenced by a variety of factors such as polymer lattice orientations, film thickness, solvent content within the film, and counterion ingress and egress into and out of the polymeric film [19, 20]. A number of electrochemical and hyphenated electrochemical techniques have been used to characterize chemically modified electrodes and these include cyclic voltammetry, square wave voltammetry, spectroelectrochemistry, electrochemical quartz crystal microbalance, and electrochemical impedance spectroscopy, just to name a few.

3. Classification of polymer modified electrodes

3.1 Loaded ionomers

These polymer coatings are polycation or polyanion ion-exchange polymers, which contain electroactive species. The ion-exchange polymer is attached to an electrode via solvent evaporation or spin coating which is then dipped into a solution containing electroactive species. The electroactive species are essentially bound to the ion-exchange polymer through electrostatic forces. One of the most popular and extensively used ion-exchange polymers is Nafion, a perfluorosulfonate ion-exchanger [9]. Other ion-exchange polymers which have been investigated are protonated forms of poly(vinylpyridine), and poly-N-methylpyrrole with immobilized poly(4-styrenesulfonate) ions as a cation-exchanger [9].

Anson and Oyama utilized protonated forms of poly(4-vinylpyridine) as cation exchange polymers to incorporate Ru^{3+}, hexacyanoferrate $(Fe(CN)_6^{3-/4-}$, and hexachloroiridate $(IrCl_6^{2-/3-})$ [10]. Maksymuk and Doblhofer used poly-N-methylpyrrole with immobilized poly(4-styrenesulfonate) ions as a cation-exchanger to incorporate $Fe(CN)_6^{3-/4-}$, $Ru(NH_3)_6^{3+/4+}$, $Eu^{3+/}Eu^{2+}$, $Co(en)_3^{3+/2+}$, and $Fe(C_2O_4)_3^{3-/4-}$ [11]. The electropolymerization of pyrrole, aniline, or N-methylpyrrole in the presence of polyanions such as poly(vinylsulfate), and poly-(stryrenesulfonate) forms an ion-exchanger, which is electroactive over a wide potential range [9]. Other workers have incorporated electroactive inorganic cations such as $Co(2,2'\text{-bipyridine})_3^{2+}$, and $Ru(2,2'\text{-bipyridine})_3^{2+}$ into polystyrene sulfonate [12]. The widely used Nafion has been utilized by Yagi and co-workers to investigate the charge transfer properties of tris(2,2′-bipyridine)Ru^{2+} [13]. The charge transport parameters are based upon two processes, namely, physical displacement of the electroactive center, and electron hopping or electron self-exchange between electroactive centers [9]. Although this method is useful in studying charge transport properties of redox active species within membranes, it suffers from a major disadvantage in that the electroactive species are not irreversibly bound to the ion-exchange matrix, and consequently leaches out into the solution.

3.2 Redox polymers

Polymer films with redox-active centers covalently bound to a redox inactive polymer organic moiety is a redox polymer. These polymers can be synthesized in large quantities, and then immobilized onto electrode surfaces by spin coating, or dip coating. One major obstacle associated with using preformed redox polymers is its low solubility in solvents. The solvent must evaporate when using spin coating, dip coating, or solvent evaporation methods in order to leave the preformed polymer on the electrode surface. The mechanism of charge transfer has been exclusively identified as electron hopping or electron self-exchange between neighboring redox centers. Such polymers only show conductivity over a limited potential range. Shgehara and co-workers incorporated $Fe(CN)_5^{3-}$ into poly(4-vinylpyridine) via coordination, and attached the polymer to graphite electrodes by solvent evaporation [21].

One of the most successful and appealing ways to form redox polymers on electrode surfaces is through electropolymerization using cyclic voltammetry. Electropolymerization is an attractive approach because it allows film thickness to be consistently reproduced, and monitored. In addition, electrode coverage is normally complete after many scans, and requires many layers because the polymer may not grow in regular layers. This gives rise to what is commonly called a multilayered polymeric thin film. As mentioned previously, these electrodes show a greater electrochemical response than monolayer coverages. Multilayered films have more electroactive sites in the form of layers and as many as 1000 layers can be formed on a single electrode surface [1, 22]. Iron and ruthenium complexes of 2,2′-bipyridine have been electropolymerized on platinum, glassy carbon, and metal oxide electrodes by Ellis and Meyer to selectively direct charge transfer to redox species in solution [23]. The same research group performed oxidative electropolymerization of ruthenium complexes of 4-aminopyridine, 3-aminopyridine, and 5-amino-1,10 phenanthroline [24]. Metalloporphyrins have been used in electropolymerization by several research groups. Derivatives of nickel(II) and cobalt(II)-tetraphenylporphyrins with amino-, hydroxy-, and pyrrole-, substituents have been electropolymerized by Murray and co-workers [25]. Such compounds are used because of their rich chemistry in terms of electron transfer properties. These type of compounds can be used as possible catalysts since the central metal ion is capable of axially coordinating with extra ligands. The mechanism(s) for generating polymeric multilayered films are based upon similar principles of either forming radical cations or anions which couple together. This process is looked into further under the discussion of "electropolymerization reactions."

3.3 Electronically conducting polymers

Electronically conducting polymers are marked by extended π-electron systems which are delocalized. These are can be formed in a similar manner as redox polymers using electropolymerization schemes. One of the first compounds to be used in forming these polymers was pyrrole by Diaz and Kanazawa [26]. Later, other research groups explored aniline, furan, and other heterocyclic aromatic compounds for electropolymerization. The polymerization of these compounds proceeds by the mechanism(s) similar to derivatives of metalloporphyrins. For example, pyrrole forms polymeric films via radical cation intermediates. In all cases, electropolymerization is governed by Faraday's Law and hence it is easy to monitor film thickness.

Charge transport within the electropolymerized film is attributed to interchain electron hopping coupled the egress or ingress of counterions to maintain charge electroneutrality during electrochemical characterization or stimulation. The chain alignment plays a significant role in the kinetics of charge transfer, and the conductivity of these films. Film thickness can be an important factor in designing sensors because analytes may penetrate through the film to the electrode surface, partially through the film, or to the film solution interface in order to undergo a charge transfer electrocatalytic reaction [11].

4. Electropolymerization reactions

Electropolymerization can be performed using cyclic voltammetry (CV) wherein experimental CV conditions such as scan rate, number of cycles, and the potential window all affect the overall composition of the resultant polymeric thin films. Electropolymerization reactions are first preceded by a small of amount of adsorption onto the electrode surface. In order for a monomer to be electropolymerized on an electrode surface, it must possess functional groups which can be oxidized or reduced via a π^* molecular orbital or non-bonding electrons on a heteroatom. The most common functional groups noted to facilitate electropolymerizations reactions are amino-, hydroxy-, and vinyl groups. In these instances, there is a lone pair of electrons or a π-electron system available. These organic moieties give rise to radical cations or anions depending upon whether oxidation or reduction is the initiation step. These radical species couple at the electrode surface, and undergo redox processes to generate the thin film. The conductivity of the developing film on the electrode surface serves to mediate charge transfer at the film-solution interface so that more monomers can be incorporated into the growing film. Statistically, increasing the number of aforementioned functional groups on a compound favors the formation of radical species and a heavily cross-linked polymer film. In addition, multiple functional groups on the monomer for initiating polymerization reactions promote the correct orientation for the incorporation of monomer units into the growing polymer film. This is significant when one considers how the electrode surface can be fully covered with up to 1000 monolayers.

4.1 Oxidative electropolymerization

Oxidative electropolymerization can be performed when compounds have such substituents such as amino-, and hydroxy- groups. It has been well-established that with such functional groups electropolymerization proceeds by oxidation involving the lone pair of electrons on the heteroatom [27]. The electrooxidation of aniline (1) in **Figure 1**, via the lone pair of electrons of the nitrogen yields the radical cation (2). The radical cation (2), loses a proton (3), and then undergoes radical coupling to form (4). The dimers are oxidized to form N=N double bonds (5).

Our research group has applied this mechanism to the electropolymerization of metal(II)-tetraaminophthalocyanine [M(II)TAPc] to prepare polymer thin films. For compounds such as M(II)TAPc multiple amine sites allow polymerization to continue, and film thickness to increase with greater crosslinking; the initiation step for generating poly-M(II)TAPc begins with an electrochemical oxidation of the amino- substituent that produces a radical cation. In the electropolymerization of zinc, and cobalt-tetraaminophthalo-cyanine shown in **Figure 2** using cyclic voltammetry, three reversible redox couples were identified [28, 29]. All electrochemical transformations within the zinc complex occur on the ligand system whereas CoTAPc has redox reactions on the ligand and central metal. The nonzero current of these polymer films at the negative switching potential is the result of the capacitance of the polymer films due to extensive crosslinking facilitated by the multiple amine sites on the compound. Metal complexes of ruthenium and iron involving the 5-amino-1-10-phenanthroline ligand shown in **Figure 3** have been electropolymerized onto glassy carbon, platinum, and ITO electrode surfaces.

These films on indium tin oxide surfaces demonstrate reversible electrochromism when the applied potential changed from 0 to 1.50 V (vs Ag/AgCl in 3 M KCl). Several research groups have shown that the electropolymerization of these complexes

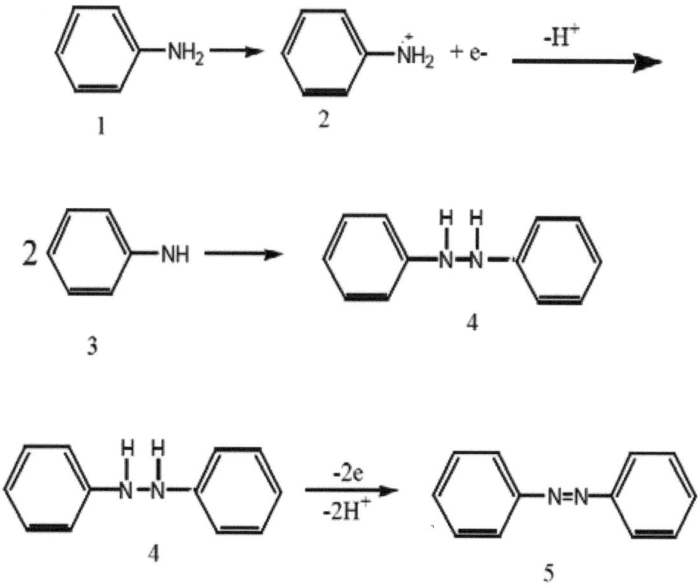

Figure 1.
Electropolymerization reactions of aniline.

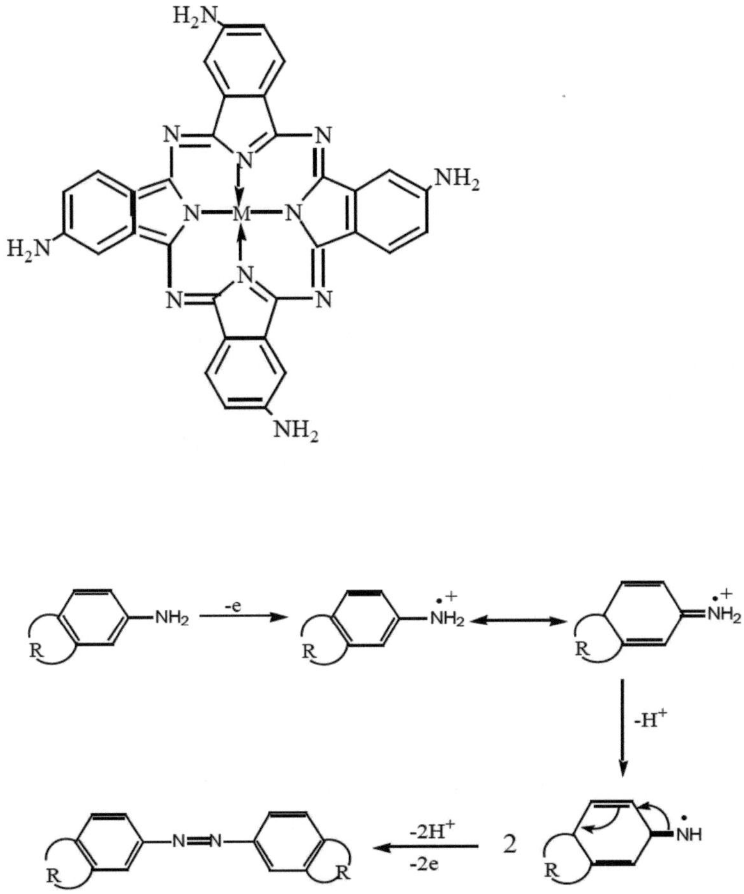

Figure 2.
Structure of metal(II)-4, 9, 16, 23-tetraaminophthalo-cyanine [M(II)TAPc] and corresponding electropolymerization reactions.

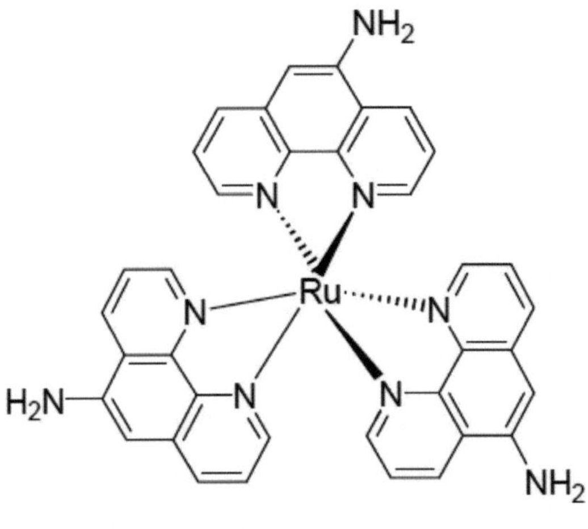

Figure 3.
Structure of 5-amino-1, 10-phenanthroline ruthenium(II).

proceeds in a similar manner as the TAPc metal complexes (**Figures 4** and **5**) but the electrochemical reversibility is more noted with the 5-amino-1-10-phenanthroline ligand.

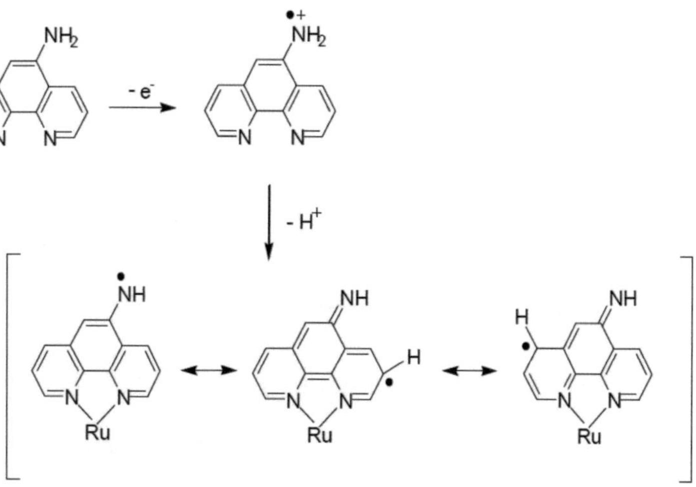

Figure 4.
Electrochemical initiation step involving the 5-amino-1, 10-phenanthroline ligand.

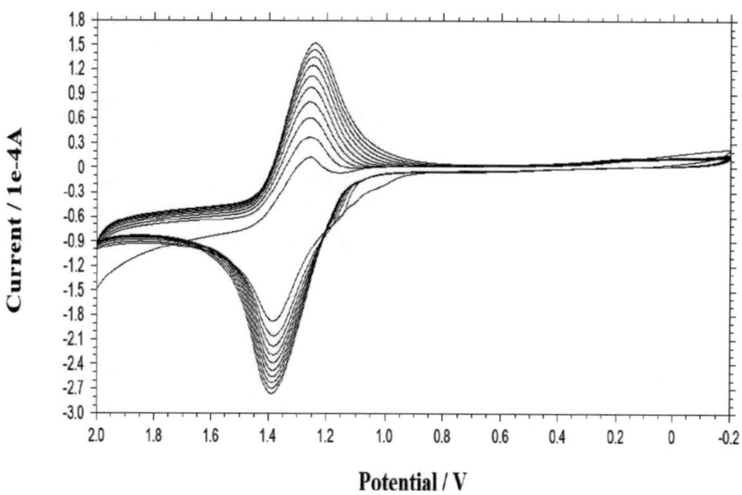

Figure 5.
Cyclic voltammogram of the electropolymerization of 5-amino-1, 10-phenanthroline iron(II).

5. Characterization of chemically modified electrodes

Chemically modified electrodes have been characterized using a variety of electrochemical, spectroscopic, mechanical, and hyphenated electrochemical-spectroscopic techniques. The most notably electrochemical characterization tools are cyclic voltammetry, chronocoulometry, quartz crystal microbalance, electrochemical impedance spectroscopy, and spectroelectrochemistry. Using a dynamic potential scanning method such as cyclic voltammetry provides both quantitative and qualitative data on the nature of electropolymerized films. For example, when characterizing polymer films using cyclic voltammetry, it has been well-established that the peak width at half-height for cathodic and anodic peaks is 90.4 mV/n, where n represents the number of electrons involved the redox transformation. However, many electropolymerized films have peak widths at half height that exceeds 90.4 mV/n (**Figure 6**). This is indicative of repulsive or attractive forces in the film or the presence of non-equivalent redox sites within the film. The electrochemical response of these films is

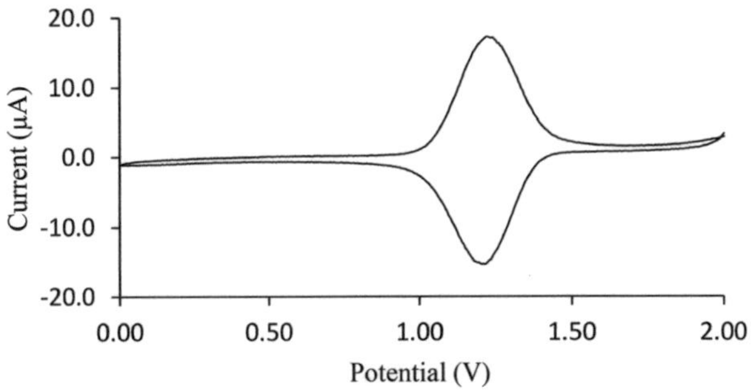

Figure 6.
Characterization of 5-amino-1, 10-phenanthroline iron(II) polymer film on a glassy carbon electrode.

the result of electron self-exchange reactions, or "electron-hopping" within the film. Concomitantly, egress or ingress of counterions from the polymer film or supporting electrolyte solution, respectively occurs when the film is oxidized or reduced using cyclic potentiodynamic conditions.

Electrochemical quartz crystal microbalance (EQCM) during CV studies have been used to determine the number of ions and associated solvent molecules entering or leaving polymer film as the films become oxidized or reduced. This provides a better understanding of the composition of polymer films real-time when they undergo a redox transformation. Cyclic voltammetry coupled with chronocoulometry has been used to determine the apparent charge transfer coefficient for electron hopping-self-exchange reactions for polymer films. Electrochemical impedance spectroscopy is widely used to study thin dynamics, corrosion of surfaces, to guide the development of sensors and biosensors, and a host of other applications.

5.1 Electrochemical quartz crystal microbalance studies

Quartz crystal microbalance (QCM) studies are based upon a deviation of the oscillation resonant frequency of a quartz crystal when the mass of the crystal changes. Oscillation frequency decreases as mass of the crystal increases. The instrumentation is sensitive enough to monitor the adsorption of gases onto the surface of the quartz crystal. When electrochemistry is combined with QCM as a hyphenated technique (EQCM), simultaneous *in situ* electrochemical and mass measurements at electrode surfaces are possible which allows electropolymerization mechanisms to be studied [30, 31]. Using EQCM it would be possible to determine the potential that causes the initiation step for electropolymerization and to determine the amount of material deposited on the electrode surface. It has been shown that reduction or oxidation processes occurring within a polymer film require that charge electroneutrality be maintained within the film. This necessitates the egress or ingress of solvated counterions with respect to the film, which can cause the film to swell.

$$\Delta f = \frac{-2f_0^2 \Delta m}{A(\mu\rho)^{\frac{1}{2}}} \tag{1}$$

$$\Delta f = \frac{(MW)(C_f)(Q)}{nF} \tag{2}$$

This impacts charge transfer within the films and may reduce the mechanical stability of the polymer film and adhesion of the polymer film to the electrode.

Research groups have used EQCM, an *in situ* mass measurement-electrochemical method, to monitor ion transport, solvent transport, and charge transfer properties within the films. The goal for these studies is oftentimes to discriminate between solvent movement and ionic movement within the films that should provide insight into whether or not charge transfer and charge electroneutrality within the films is limited by counterion movement within the films. The relationship between the change in frequency (Δf) and the change in mass (Δm) of a coated quartz crystal is given by the Sauerbrey equation (Eq. (1)) [32], where μ and ρ are the shear modulus and density of the quartz crystal, respectively. Eq. (2) is used to determine the apparent molar mass of the species causing the mass change where C_f is the mass sensitivity constant for the crystal. The applicability of this equation for solutions requires a uniform distribution of material across the surface of the crystal. In addition, the film must be rigid enough at all oxidation states to avoid dampening the shear wave that is applied to the crystal (i.e., the coating does represent a viscous medium) [30, 31]. First, in order to validate the Sauerbrey equation, films of various thicknesses based on the monomers are prepared. It is beneficial to use Atomic Force Microscopy (AFM) to verify that the material is uniformly distributed on the surface. Establishing a linear correlation between film thickness and Δf allow researchers to determine the rigidity of the film, and validate the applicability of the Sauerbrey Equation [32]. Based upon Eqs. (1), (2), it is possible to determine the molar mass, and hence, the species causing the change in mass of the film. For each cyclic voltammetric characterization of polymer film, peak separation values of the redox couples, with respect to scan rate and number of electropolymerization cycles can be evaluated to determine polymer film dynamics and characteristics. At low scan rates of cyclic voltammetric characterization, peak separation values should be near 0 V for facile charge transfer of materials immobilized onto electrode surfaces, which indicates Nernstian equilibrium with a applied potential. However, a constant low scan rate of characterization, with increasing film thickness may increase the peak separation due to limitations of charge transfer, ionic mobility, or polymer lattice motions necessary to accommodate charge transfer. Atomic force microscopy and SEM measurements provides information on changes in polymer structure after cyclic voltammetric characterization of the films and can be compared to EQCM measurements.

5.2 Determination of the apparent charge transfer diffusion coefficient

For the electrochemical characterization of the polymer films, the apparent charge transfer diffusion coefficient for a polymer thin film of a given thickness can be determined through chronoamperometric-coulometric methods, using the Cottrell equation (Eq. (3)) [28].

$$I=\frac{nFAD_{ct}^{\frac{1}{2}}C\pi^{\frac{1}{2}}}{\sqrt{t}} \tag{3}$$

The current response for an electroactive compound is given by the Cottrell equation where C and D_{ct} are the concentration (mol cm^{-3}) of redox species and apparent charge transfer diffusion coefficient (cm^2 s^{-1}), respectively. The other symbols have their normal electrochemical convention. The slope of the linear plot, I versus $t^{-1/2}$, allows the product $D_{ct}^{1/2}C$ to be calculated. Recently, we have shown that AFM can be used to determine film thicknesses and surface roughness factors for both redox polymer films and organic conducting polymer films and these properties change as the number of electropolymerization cycles change. Using the film thickness, the concentration of redox species and hence D_{ct} can be determined. The product of $D_{ct}^{1/2}C$ can also be evaluated from CV characterization data. If the mechanism proposed by Engler and Kaufman is followed for these systems, the redox sites in metal containing films that are near the outer boundary at the solution-film interface are reduced and oxidized via electron transfer between neighboring redox centers [19, 33]. The value of D_{ct} is then translated to the rate of electron transfer which may be influenced by counterion ingress and egress from film, solvent content within the film, and polymer lattice orientations, any of which could be rate determining for the electron transfer. Interpretation of D_{ct} values for the polymer films provides information on the maximum limits for electron transfer rates with an electroactive analyte in solution.

5.3 Rotating disk electrode voltammetry

Several possibilities exist for electron transfer between the analyte in solution and the polymer film and will affect the reaction between the redox mediator and the respective analyte. The analyte may penetrate to the electrode-film interface, several monolayers into the film, or interact with the film-solution interface. Koutecky-Levich plots [34] can be obtained from the polymer films on rotating disk electrodes at various rotational velocities (ω), and concentrations (C_h) of analyte (i.e., glucose, hydrazine, and metal ions). Analyzing variable film thicknesses helps determine if electron self-exchange in the polymer film or diffusion of analyte through the polymer film is rate limiting. This helps elucidate the kinetics of analyte reduction or oxidation. The overall current in the RDE method is composed of a catalytic current (I_{cat}), kinetic current (I_k), and Levich current (I_{lev}), as shown in Eqs. (4)–(6). The catalytic current is derived from the mediated charge transfer reaction, the kinetic current from the cross exchange reaction between the polymer film and analyte, and the Levich current from the mass transfer of analyte in solution [34].

$$\frac{1}{I_{cat}} = \frac{1}{I_{lev}} + \frac{1}{I_k} \tag{4}$$

$$I_{lev} = 0.620nFAD^{\frac{2}{3}}\omega^{\frac{1}{2}}v^{\frac{-1}{6}}C_h \tag{5}$$

$$I_k = nFAk_hC_h \tag{6}$$

Extrapolation of the linear plot of $1/I_{cat}$ with respect to $1/\omega^{1/2}$ should give an intercept of $1/I_k$ when extrapolated to $\omega^{-1/2} = 0$ to determine the rate constant, k_h, for the cross reaction between the analyte and polymer film. For these experiments, films of different thicknesses prepared on electrode surfaces and the currents from the RDE measurements helps determine if the current response is diffusion limited for the analyte.

5.4 Electrochemical impedance spectroscopy

Electrochemical impedance spectroscopy (EIS) has been used by a variety of researchers to study charge transfer processes at electrode/solution interfaces, ionic transport within films, and diffusion mass transport processes [35]. Corrosion is

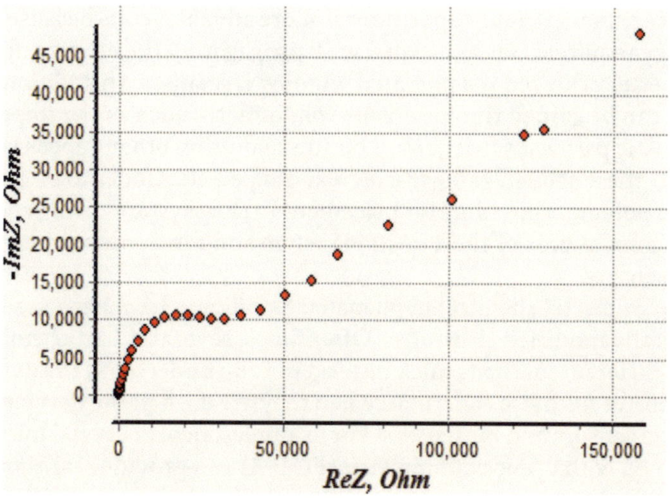

Figure 7.
Electrochemical impedance spectroscopy of electropolymerized film of 5-amino-1, 10-phenathroline-iron(II) on a glassy carbon electrode.

another area of intense research and makes extensive use of EIS measurements [36]. Using EIS it is possible to detect flaws in coatings on metals and to determine when a coating on a metal and alloy surface fails. Electrochemical impedance measurements are used to better understand how sensors operate theoretically by building equivalent circuits for charge transfer processes, mass transport processes, and ionic migration. Some films within electrochemical sensors become porous due to analyte diffusion and/or migration effects after extended use and EIS can be used to evaluate changes in morphology of the films that in turn affect sensor and biosensor performance. Electrochemical impedance spectroscopy employs alternating circuit (AC) theory to electrochemical cells [37, 38]. That is, an electrochemical cell can be modeled with electrical components consisting of resistors, capacitors, and inductors in a series or parallel arrangement. However, in complex systems, the arrangement of the equivalent circuit design is generally a series–parallel combination arrangement. Electron flow in an electrochemical sensor can be hindered via slow heterogeneous or homogeneous kinetics, slow diffusional or migration processes, or morphological changes in the films used. This total impedance is the combination of resistors, capacitors, and inductors. In EIS, an AC signal of approximately 5–20 mV is superimposed on a DC component of the electrochemical system that provides a pseudo-linear response over a wide frequency range. Films characterized with EIS ideally show charge transfer resistance (Rct), which is given by a semicircle and the semicircle intersects the real-Z' axis at high frequencies. At lower frequencies, when time is allowed for diffusion effects to be realized, a Warburg impedance dominates the charge transfer process as shown in **Figure 7**. Impedance measurements will also provide information about the porosity of films before and after use by determining the pore impedance (Z_{pore}). A time profile of Z_{pore} will indicate: (1) formation of pores, (2) collapsing of pores, or (3) constant pore structure [37].

6. Conclusion

An effective method of altering the properties of an electrode surface to prepare chemically modified electrodes is through the use of electropolymerization reactions. Electropolymerization by cyclic voltammetry as opposed to constant potential or constant current depositions is more advantageous because there are certain energy requirements associated with propagating the electropolymerization that can be better achieved if a potential window is scanned. In addition, much information can be gained through cyclic voltammetry such as electropolymerization mechanistic pathways that cannot be attained using other deposition methods. Furthermore, these schemes allows irregular-shaped electrodes to be effectively covered with polymer thin films for placement in flow systems using chemically modified systems as part of the detection system and for the development of sensors and biosensors.

A number of electrochemical hyphenated techniques have been used to characterize chemically modified electrodes. Oftentimes, several electrochemical experiments using different methods must be used to fully understand the dynamics of charge transfer for these electrodes when they are used in the development of amperometric sensors and biosensors. For examples, quartz crystal microbalance when combined with cyclic voltammetry (EQCM) can provide real-time information about mass changes during electrochemical stimulation of chemically modified electrodes. One goal of electrochemical characterization of thin films is to correlate the properties of the polymer films to the processes occurring during the electropolymerization when the polymer film is prepared. One technique, electrochemical impedance spectroscopy, which is typically used to understand corrosion has been used to provide a clearer picture of the factors limiting charge transfer processes at the chemical modified electrode surfaces. Using this method, one experiment has the capability to give information equivalent to several experiments in cyclic voltammetry and chronocoulometry. However, no one electrochemical technique can provide all the information necessary to understand the dynamics of electron transfer phenomena within or at chemically modified electrode surfaces.

Acknowledgements

Funding for this work and research has been provided by the Hope College Chemistry Department, the National Science Foundation (Grant #1263097), and the Schaap Endowed Fund for Undergraduate Research

Author details

Kenneth L. Brown
Hope College, Holland, MI, USA

*Address all correspondence to: brownk@hope.edu

References

[1] Murray R. Chemically modified electrodes. Accounts of Chemical Research. 1980;**13**:135-141

[2] Hillman AR. Electrochemical Science and Technology of Polymers. Elsevier: Amsterdam; 1987

[3] Hart JP, Wring SA. Chemically modified, carbon-based electrodes and their application as electrochemical sensors for the analysis of biologically important compounds. A review. Analyst. 1992;**117**:1215

[4] Kaneko M, Wohrle O. Polymer-coated electrodes: New materials for science and industry. Advances in Polymer Science. 1988;**84**:140

[5] Tse Y-H, Janda P, Lever ABP. Electrode with electrochemically deposited N,N',N",N"'-tetramethyltetra-3,4-pyridinoporphyrazinocobalt(I) for detection of sulfide ion. Analytical Chemistry. 1992;**66**:384-390

[6] Wang J. Electrocatalytic reduction and flow injection analysis of organic peroxides at polymeric tetra-amino iron phthalocyanine modified electrode. Analytical Letters. 1996;**29**:1575-1587

[7] Schachl L, Alemu H, Klacher K, Jezkova J. Flow injection determination of hydrogen peroxide using a carbon paste electrode modified with a manganese dioxide film. Analytical Letters. 1997;**30**:2655-2673

[8] Collins G, Rose-Pehrsson SL. Chemiluminescent chemical sensors for oxygen and nitrogen dioxide. Analytical Chemistry. 1995;**67**:2224-2230

[9] Lane RF, Hubbard AT. Electrochemistry of chemisorbed molecules. I. Reactants connected to electrodes through olefinic substituents. Journal of Physical Chemistry. 1973;**77**:1401

[10] Oyama N, Anson FC. Factors affecting the electrochemical responses of metal complexes at pyrolytic graphite electrodes coated with films of poly(4-Vinylpyridine). Journal of Electroanalytical Chemistry. 1980;**127**:640-647

[11] Maksymiuk L, Doblhofer K. Kinetics and mechanism of charge-transfer reactions between conducting polymers and redox ions in electrolytes. Electrochimica Acta. 1994;**39**:217-227

[12] Shaw BR, Haight GP, Faulkner LR. Electrochemical preparation of electrodes coated with crosslinked poly(styrenesulfonate). Journal of Electroanalytical Chemistry. 1982;**140**:147-153

[13] Yagi M, Mitsumoto T, Kaneko M. Potential-step chronocoulospectrometry of a polymer membrane incorporating tris(2,2′-bipyridine) ruthenium (II) complex. Journal of Electroanalytical Chemistry. 1997;**437**:219-223

[14] Calvert JM, Schmehl RH, Sullivan PB, Facci JS, Meyer TJ, Murray RW. Synthetic and mechanistic investigations of the reductive electrochemical polymerization of vinyl-containing complexes of iron(II), ruthenium(II), and osmium(II), Inorganic Chemistry. 1983;**22**:2151-2162

[15] Kobayashi T, Yoneyama H, Tamura H. Polyaniline film-coated electrodes as electrochromic display devices. Journal of Electroanalytical Chemistry. 1984:419-423

[16] Lyons MG. Electroactive Polymer Electrochemistry. New York, NY: Plenum Press; 1994

[17] Feldberg SW. Reinterpretation of polypyrrole electrochemistry. Consideration of capactive currents in redox switching of conducing polymers. Journal of the American Chemical Society. 1984;**106**:4671-4674

[18] Walczak RM, Cowart JS, Reynolds JR. Tethered PProDOTs: Conformationally restricted 3,4-propylenedioxythiophene based electroactive polymers. Journal of Materials Chemistry. 2007;**17**:254-260

[19] Sotzing GA, Reynolds JR. Electrochromic conducting polymers via electrochemical polymerization of bis(2-(3,4-ethylenedioxy)thienyl) monomers. Chemistry of Materials. 1996;**8**:882-889

[20] Paulson SC, Sapp SA, Elliott CM. Electrochemical and spectroelectrochemical investigations into the nature of charge-trapping in electrochemically generated homopolymer films of tris(4-vinyl-4'-methyl-2,2'-bipyridine)ruthenium(II). The Journal of Physical Chemistry B. 2001;**105**:8718-8724

[21] Shgehara L, Oyama N, Anson FC. Electrochemical responses of electrodes coated with redox polymers. Evidence for control of charge-transfer rates across polymeric layers by electron exchange between incorporated redox sites. Journal of the American Chemical Society. 1981;**103**:2552-2558

[22] Nyasulu FWW, Mottola HA. Electrochemical behavior of 5-amino-1,10-phenanthroline and oxidative electropolymerization of tris[5-amino-1,10-phenanthroline]iron(II). Journal of Electroanalytical Chemistry. 1988;**239**:175-186

[23] Ellis CD, Murphy WR, Meyer TJ. Selectivity and directed charge transfer through an electroactive metallopolymer film. Journal of the American Chemical Society. 1981;**103**:8480-7483

[24] Ellis CD, Murray RW, Meyer TJ, Margerur LD. Oxidative electropolymerization of polypyridyl complexes of ruthenium. Inorganic Chemistry. 1983;**22**:1283-1291

[25] Bettelheim A, White BA, Raybuck SA, Murray RW. Electrochemical polymerization of amino-, pyrrole-, and hydroxy-substituted tetraphenylporphyrins. Inorganic Chemistry. 1987;**26**:1009-1017

[26] Diaz AF, Kanazawa KK, Gardini GP. Electrochemical polymerization of pyrrole. Journal of the Chemical Society, Chemical Communications. 1979:635

[27] Genies AF, Kanazawa KK, Gardini GP. Spectroelectrochemical study of polypyrrole films. Journal of Electroanalytical Chemistry. 1983;**149**:101-113

[28] Brown KL, Mottola HA. Voltammetric, Chronocoulometric, and spectroelectrochemical Studies of Electropolymerized Films Based on Cu(II/I)- 4, 9,16, 23-Tetraaminophthalocyanine. Langmuir. 1998;**14**:3411-3417

[29] Brown KL, Shaw J, Ambrose M, Mottola HA. Voltammetric, Chronocoulometric and Spectroelectrochemical Studies of Electropolymerized Films Based on Co(III/II)-and Zn(II)-4, 9, 16, 23-Tetraaminophthalocyanine: Effect of High pH. Microchemical Journal. 2002;**72**:285-298

[30] Tjarnhage T, Sharp M. The use of a quartz crystal microbalance combined with ellipsometry and cyclic voltammetry for determining some basic characteristics of an electroactive

polymer film. Electrochimica Acta. 1994;**39**(5):623-628

[31] Cui S, Park S. Electrochemistry of conductive polymers XXIII: Polyaniline growth studied by electrochemical quartz crystal microbalance measurements. Synthetic Metals. 1999;**105**:91-98

[32] French HM, Henderson MJ, Hillman AR, Vieil E. Ion and solvent transfer discrimination at a nickel hydroxide film exposed to LiOH by combined electrochemical quartz crystal microbalance (EQCM) and probe beam deflection (PBD) techniques. Journal of Electroanalytical Chemistry. 2001;**500**:192-207

[33] Kaufman FB, Schroeder AH, Engler EM, Kramer SR, Chambers JQ. Ion and electron transport in stable, electroactive tetrathiafulvalene polymer coated electrodes. Journal of the American Chemical Society. 1980;**102**(2):483-488

[34] Maksymuk K, Doblhofer K. Kinetics and mechanism of charge-transfer reactions between conducting polymers and redox ions in electrolytes. Electrochemica Acta. 1994;**39**(2):217-227

[35] Mansfeld F. Use of electrochemical impedance spectroscopy for the study of corrosion protection by polymer coating. Journal of Applied Electrochemistry. 1995;**25**:187-202

[36] Zelinka SL, Stone DS, Rammer DR. Electrochemical corrosion testing of fasteners in extracts of treated wood. Corrosion Science. 2008;**50**(5):1251-1257

[37] Lvovich VF. Impedance Spectroscopy: Applications to Electrochemical and Dielectric Phenomena. Hoboken, New Jersey: John Wiley and Sons: 2012

[38] Krause S. Encyclopedia of Interfacial Chemistry: Surface Science and Electrochemistry. Vol. 3. Wiley-VCH; 2001

Chapter 5

Investigation of Electrochemical Pitting Corrosion by Linear Sweep Voltammetry: A Fast and Robust Approach

Shashanka Rajendrachari

Abstract

Generally, impedance spectroscopy, cyclic voltammetry and polarographic methods are used to study the pitting corrosion of steel, stainless steel and many different alloys. But one can also use linear sweep voltammetry (LSV) to investigate the pitting corrosion phenomenon. LSV is having many advantages over other traditional methods; but more research should take place in this area to foreshorten the lacuna. It is an important electrochemical method that involves solid electrode, fixed potential and fast scan rate. The advantage of using LSV in determining the pitting corrosion is less time required in the order of few seconds, and there is no need of keeping the samples in NaCl or any other electrolytes for many months.

Keywords: linear sweep voltammetry, pitting corrosion, stainless steel, electrolyte, scan rate

1. Introduction

Over the few decades, the use of stainless steel has been increased tremendously in various fields due to its beneficial properties like high corrosion resistance, low thermal expansion, high energy absorption, high strength, good weldability and high toughness [1]. The ferritic stainless steel is composed of a lesser quantity of expensive Ni and 10–20 wt.% of Cr, and it has body-centred cubic (BCC) lattice structure [2]. The properties like high thermal conductivity, low thermal expansion, wear and creep resistance, higher yield strength, excellent high-temperature oxidation resistance and less stress corrosion properties had made the ferritic stainless steel one of the important grades of stainless steel [3]. Some of its applications of ferritic stainless steel are fabricating cold water tank, refrigeration cabinets, bench work, chemical and food processing, water treatment plant, street furniture, electrical cabinets, etc. [4].

However, duplex stainless steel is a combination of equal proportions of austenitic and ferritic stainless steel grades. Ferrite phase imparts high strength, while austenite contributes the toughness and high corrosion resistance [5]. Duplex stainless steel possesses amalgamated properties of both ferritic and austenitic stainless steel, and hence it is one of the popular and widely applicable stainless steels. Duplex stainless steels have a wide range of applications in various fields like chemical, oil, petrochemical, marine, nuclear power to paper and pulp industries [6–8].

Pitting corrosion is a localised hastened dissolution of metal due to the destruction of the protective passive film on the metal surface. The important steps involved in pitting process are breakdown of the passive film, metastable pitting and pit growth. The mechanism of pitting corrosion involves the dissolution of protective passive film and gradual acidification of the electrolyte caused by its insufficient aeration [9]. This increases the pH of the pits by increasing the anion concentration. Most of the engineering metals and alloys are useful only because of passive films (thickness of nano to micron range) and naturally forming oxide layers on the metal surfaces. This will greatly reduce the rate of corrosion of the metals as well as alloys [10]. But these types of passive films are more often amenable to localised destruction resulting in accelerated dissolution of the underlying metal. This type of localised pitting corrosion results to the accelerated failure of structural components by perforation or pitting or by acting as an initiation site for cracking. Many researchers all over the world published many research papers on pitting corrosion of stainless steel by different electrochemical methods.

Shankar et al. [11] studied pitting corrosion resistance of yttria-dispersed stainless steel by cyclic polarisation experiments in 3.56 wt.% NaCl solution. They concluded that the addition of Y2O3 did not affect the pitting corrosion resistance. Ningshen et al. [12] reported the corrosion resistance of 12 and 15% Cr oxide dispersion strengthened (ODS) steels in 3 and 9 M HNO_3, respectively. They observed that 12% chromium ODS steel exhibits higher corrosion rate than 15% chromium ODS steel at both 3 and 9 M HNO_3 concentrations. Balaji et al. [13] studied the corrosion resistance of yttria aluminium garnet (YAG)-dispersed austenitic stainless steel at different concentrations (1, 2.5 and 7.5 wt.%) of YAG. The corrosion studies were carried out in 0.1 N H_2SO_4 using potentiodynamic polarisation, and they reported that addition of YAG does not increase corrosion rate appreciably, but super-solidus sintering shows higher corrosion resistance than solid-state sintering.

Most of the corrosion studies were carried out using electrochemical methods such as impedance spectroscopy [14], polarographic methods [15], cyclic voltammetry [16], etc. But very limited literature is available so far on the corrosion study of stainless steel samples by linear sweep voltammetry method. Linear sweep voltammetry (LSV) is one of the most important methods of electroanalytical chemistry [17–19], initiated by Heyrovsky. He was honoured with a Nobel Prize in Chemistry in the year 1959 due to his pioneering work on cyclic voltammetry and linear sweep voltammetry. LSV is an important voltammetric method in which the current at a working electrode is measured, while the potential between the working electrode and a reference electrode is swept linearly with time [20]. The Randles-Sevcik theory is mainly based on an assumption of diffusion limitation of the active species in a neutral liquid electrolyte, driven by fast reactions at the working electrode [21–23]. In the case of LSV, a fixed potential range is applied similar to the potential step measurements.

The scan rate (v) is calculated from the slope of the line. We can easily alter the scan rate by changing the time required for sweep. The characteristics of the linear sweep voltammogram mainly depend on the following factors:

- The voltage scan rate
- The rate of the electron transfer reaction
- The chemical reactivity of the electroactive species

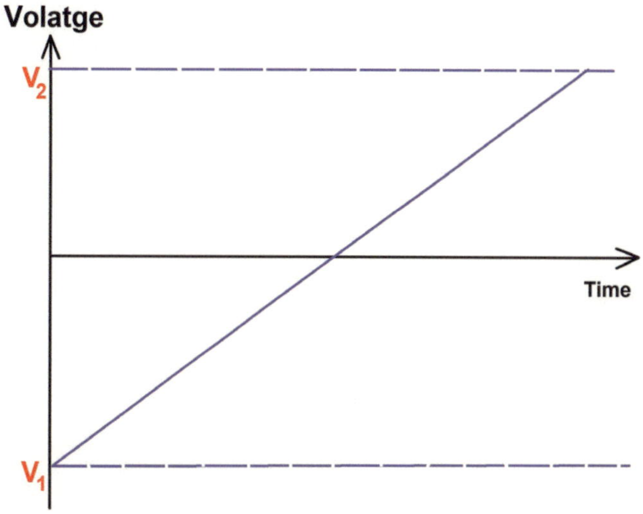

Figure 1.
Linear sweep voltammetry curve [24].

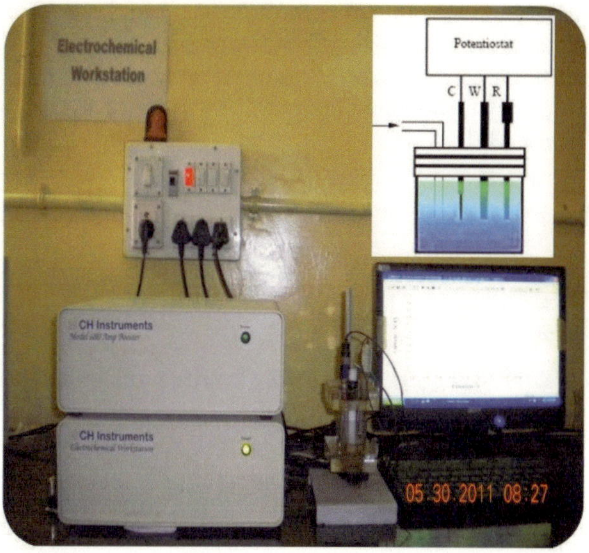

Figure 2.
Linear cyclic voltammetry experimental set-up [25].

But, however, in LSV, the voltage is scanned from a lower limit to an upper limit as depicted below (**Figures 1** and **2**).

This chapter provides an overview of the factors influencing the pitting corrosion of duplex and ferritic stainless steel samples fabricated by spark plasma sintering. The detailed information of fabrication, characterisation and consolidation of duplex and ferritic stainless steels was published by the author in his previous research articles [26–37]. Both the types of stainless steels are used in different fields, and we need to study their corrosion properties before using them as engineering materials. The author has explained the effect of composition, time and different concentrations of electrolytes on the pitting corrosion of duplex and ferritic stainless steels by using LSV.

The experimental data explained in this chapter is performed by the author himself, and part of it is published elsewhere [38, 39]. The corrosion studies were carried out in a well-established three-electrode electrochemical cell using an electrochemical work station CHI-660c model by LSV method. Potential scans were collected in a freely aerated NaCl and H_2SO_4 solutions at room temperature. The experiments were carried out in an electrochemical cell containing Ag/AgCl-saturated KCl as reference electrode and stainless steel samples as working electrode (20 mm diameter) and platinum counter electrode. Corrosion studies were carried out in 0.5, 1 and 2 M concentration of NaCl and H_2SO_4 solutions at different quiet times of 2, 4, 6, 8 and 10 s by LSV method.

2. Mechanism of pitting corrosion in stainless steel

In other words, pitting corrosion is a process of depleting passive layer of the stainless steel aroused by an electrolyte rich in chloride and/or sulphides [25]. **Figure 3** depicts the mechanism of pitting corrosion process in stainless steel. There is a formation of protective Cr_2O_3 passive layer. But in the presence of anions and proper voltage, the passive layer breaks, and the voltage required to deplete the passive layer is called as pitting voltage. After breaking of Cr_2O_3 layer, initiation of pit starts. This increases the anion concentration (Cl^-) in the electrolyte and pit grows further. But in some cases, re-passivation of pit takes place, and this partially improves the corrosion-resistant properties of stainless steel.

Some of the reactions responsible for corrosion in stainless steel at NaCl electrolytes are shown below [38]:

$$\text{Anodic reaction: } Fe \rightarrow Fe^{2+} + 2e^- \text{ (dissolution of iron)}$$

$$Fe^{2+} + 2H_2O \rightarrow FeOH^+ + H_3O^+$$

Formation of $FeOH^+$ is mainly responsible for the sudden increase in current due to the dissolution of Fe metal:

$$\text{Cathodic reaction: } O_2 + 2H_2O + 4e^- \rightarrow 4OH^-$$

$$\text{Main reaction: } Fe^{2+} + 2Cl^- \rightarrow FeCl_2$$

$$FeCl_2 + 2H_2O \rightarrow Fe(OH)_2 + 2HCl$$

Formation of $Fe(OH)_2$ increases the pH of the electrolyte inside a pit from 6 to 2, which induces further corrosion process.

But in the case of H_2SO_4, re-passivation of passive layer takes place due to the following reactions [38]:

$$\text{Anodic reaction:} Fe \rightarrow Fe^{2+} + 2e^- \text{ (dissolution of iron)}$$

$$\text{Cathodic reaction: } 2H^+ + 2e^- \rightarrow H_2$$

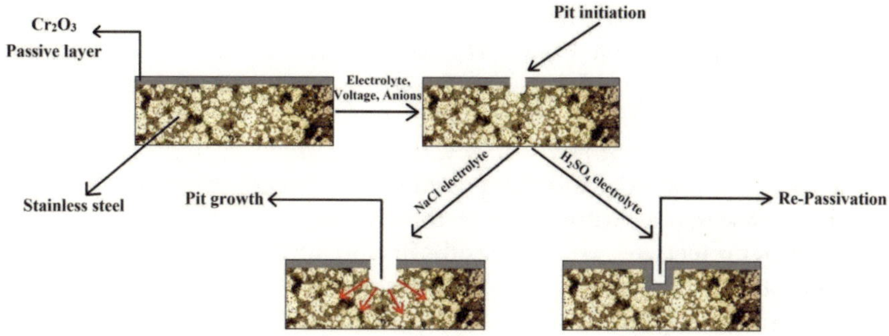

Figure 3.
Mechanism of pitting corrosion in stainless steel samples [38].

The main corrosion reaction gives the products iron sulphate and hydrogen gas as shown below:

$$\text{Main reaction: } Fe + H_2SO_4 \rightarrow FeSO_4 + H_2$$

There is a formation of $FeSO_4$ thin layer on the stainless steel surface, which acts as a passive protective layer for corrosion. But the liberated hydrogen gas scrubs off the $FeSO_4$ layer and causes corrosion [40].

3. A corrosion study of duplex and ferritic stainless steels by LSV

3.1 Different concentrations of NaCl electrolyte solution

We studied the effect of reaction time (quiet time) and different concentrations of NaCl electrolyte on pitting corrosion. The concentrations 0.5, 1 and 2 M of NaCl were prepared in double-distilled water and were used to study the pitting corrosion. The spark plasma-sintered duplex and ferritic stainless steel samples were polished to 4/0 grade finish and cleaned with distilled water before the experiment. The stainless steel whose pitting corrosion properties to be studied was kept inside the electrochemical cell containing NaCl electrolyte, counter electrode and reference electrode. LSV was performed by sweeping a potential from 0.9 to 0 V (adjusted according to the pitting potential) with different quiet times of 2, 4, 6, 8

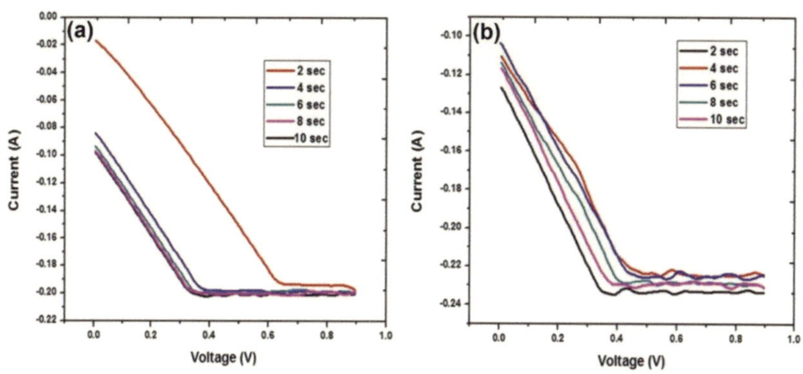

Figure 4.
Potentiometric graphs of (a) duplex stainless steel and (b) ferritic stainless steel, respectively, at 0.5 M NaCl solution [30].

and 10 s. A curve of potential and current was obtained for each individual quiet time at specific concentration. **Figure 4(a)** and **(b)** depicts the LSV curve of current versus voltage variation of duplex and ferritic stainless steel samples at 0.5 M NaCl concentrations in different quiet times.

As the potential sweeps from 0.9 to 0 V, the sharp increase in the current takes place at a particular potential and that potential is called as pitting potential (EP). The sharp increase in the current is due to the availability of more electrons after depleting Cr_2O_3 protective layer. This results in a pit, and it will grow in size if the metal is unprotected and leads to pitting corrosion. EP values of duplex and ferritic stainless steel samples were found to be 0.63 and 0.57 V, respectively. Duplex stainless steel shows more EP value than ferritic stainless steel due to more amount of Cr in duplex than ferritic stainless steel, which imparts maximum strength to interfacial bonding and forms strong oxide layer. Hence, more potential is required to break the oxide layer; the higher the pitting potential, the better is the pitting corrosion resistance [41].

Figure 5(a) and **(b)** shows the current versus voltage graphs of duplex and ferritic stainless steel samples at 1 M NaCl concentration. EP values of duplex and ferritic stainless steel samples are found to be 57 and 0.19 V, respectively. Similarly, **Figure 6(a)** and **(b)** represents the current versus voltage graphs of duplex and ferritic stainless steel samples at 2 M NaCl concentration. The duplex and ferritic stainless steel samples at 2 M NaCl show the EP value of 0.24 and 0.18 V, respectively. From the graphs it is clear that as the concentration of NaCl electrolyte increases from 0.5 to 2 M, then pitting potential of duplex and ferritic stainless steel samples decreases due to the accelerated rate of corrosion reactions at higher concentrations.

3.2 Different concentrations of H_2SO_4 electrolyte solution

The pitting corrosion studies were carried out in a same electrochemical experimental set-up with same experimental condition as the corrosion studies conducted for NaCl electrolyte. But we have used H_2SO_4 here instead of NaCl to study the effect of acid electrolyte on corrosion of duplex and ferritic stainless steel samples. We prepared 0.5, 1 and 2 M concentration of H_2SO_4 electrolyte in double-distilled water and used to study the pitting corrosion of duplex and ferritic stainless steel samples.

A sweep potential of 0.6–0 V (adjusted according to the pitting potential) was applied in LSV with different quiet times of 2, 4, 6, 8 and 10 s, respectively. A voltammetric curve was collected for each individual quiet time at a particular

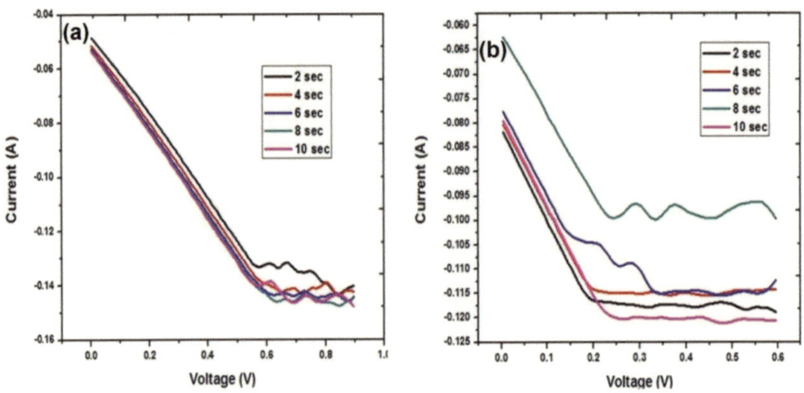

Figure 5.
Potentiometric graphs of (a) duplex stainless steel and (b) ferritic stainless steel, respectively, at 1 M NaCl solution [30].

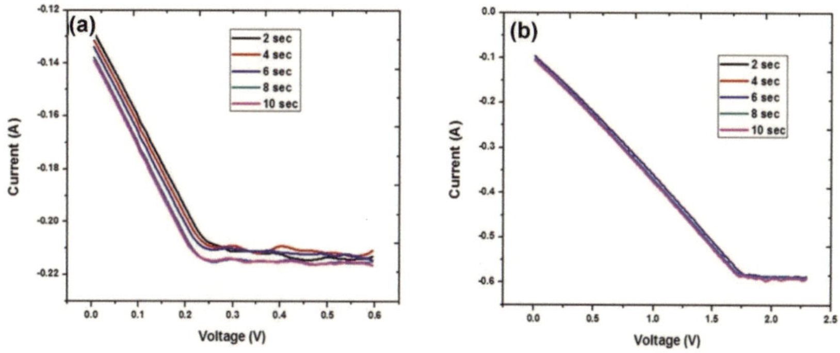

Figure 6.
Potentiometric graphs of (a) duplex stainless steel and (b) ferritic stainless steel, respectively, at 2 M NaCl solution [30].

constant concentration. LSV curve of current versus voltage variation in duplex and ferritic stainless steel samples at 0.5 M H_2SO_4 electrolyte after 2, 4, 6, 8 and 10 s are shown in **Figure 7(a)** and **(b)**, respectively. There is a sharp and sudden increase in the current between the potential 0.6–0 V similar to NaCl electrolyte. EP values of duplex and ferritic stainless steel samples are found to be 0.18 and 0.14 V, respectively.

We also studied the effect of EP at 1 and 2 M H_2SO_4 concentrations by maintaining the same procedure as explained above. **Figure 8(a)** and **(b)** represents the current versus voltage graphs of duplex and ferritic stainless steel samples at 1 M H_2SO_4 solution. EP values of duplex and ferritic stainless steel samples were found to be 0.17 and 0.14 V, respectively.

Similarly, **Figure 9(a)** and **(b)** shows the current versus voltage graphs of duplex and ferritic stainless steel samples at 2 M H_2SO_4 solution. Duplex and ferritic stainless steel samples have an EP value of 0.028 and −0.013 V, respectively, as shown in **Figure 9**.

In the pitting corrosion study during 0.5 and 1 M H_2SO_4 electrolyte, the formed protective $FeSO_4$ layer bounds strongly to the surface of both the stainless steels along with Cr_2O_3 passive layer, and hence hydrogen gas liberated during these concentrations is not enough to break the oxide layer to form a pit and to initiate corrosion. The pitting corrosion studies at 0.5 and 1 M H_2SO_4 solution concluded with higher pitting potential. But with 2 M H_2SO_4 solution, the pitting potential is very low, as the hydrogen gas liberated is sufficient to scrub off the formed $FeSO_4$

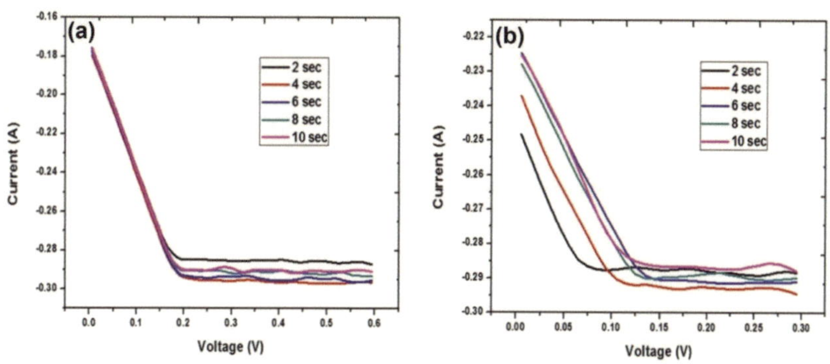

Figure 7.
Potentiometric graphs of (a) duplex stainless steel and (b) ferritic stainless steel, respectively, at 0.5 M H_2SO_4 solution [30].

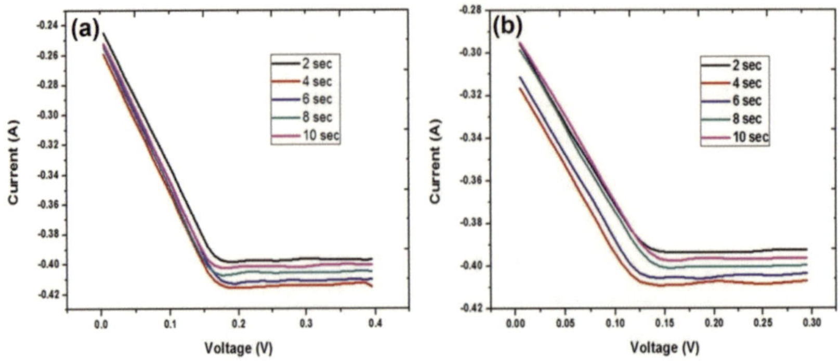

Figure 8.
Potentiometric graphs of (a) duplex stainless steel and (b) ferritic stainless steel, respectively, at 1 M H_2SO_4 solution [30].

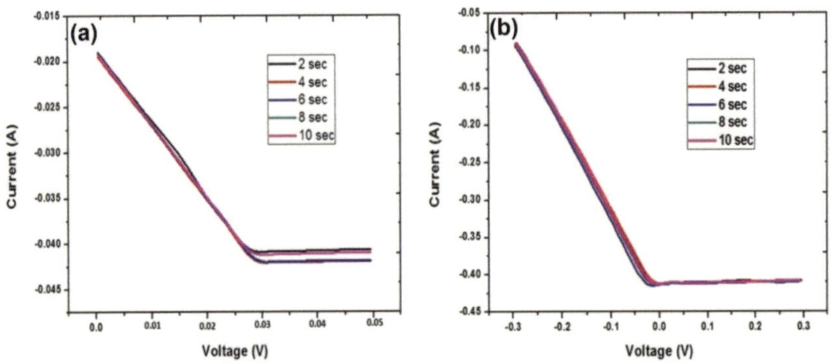

Figure 9.
Potentiometric graphs of (a) duplex stainless steel and (b) ferritic stainless steel, respectively, at 2 M H_2SO_4 solution [30].

Type of stainless steel	Concentration of the electrolyte (M)	Pitting potential (E_P) in NaCl	Pitting potential (E_P) in H_2SO_4
Duplex stainless steel	0.5	0.63	0.18
	1	0.57	0.16
	2	0.24	0.02
Ferritic stainless steel	0.5	0.57	0.14
	1	0.19	0.14
	2	0.18	−0.01

Table 1.
The EP values of duplex and ferritic stainless steels at NaCl and H_2SO_4 solutions.

layer at very low potential. As a result of this, both the stainless steel samples show low pitting potential at 2 M H_2SO_4.

Pitting potential value obtained for H_2SO_4 electrolyte is very low compared to the results obtained for NaCl electrolyte; therefore, duplex and ferritic stainless steel samples undergo corrosion easily in the presence of H_2SO_4 compared to NaCl electrolyte. The values of EP and current density with different electrolytes were tabulated in **Table 1**.

4. Post-corrosion microstructural analysis

Field emission scanning electron microscope (FESEM) images of duplex and ferritic stainless steel samples are shown in **Figure 10**. The grey-coloured regions in the FESEM images are due to pitting corrosion.

To study the further characteristics of pitting corrosion, we have used optical microscope. The optical image analysis was performed to investigate the microstructure of pitting-corroded duplex and ferritic stainless steels. **Figures 11** and **12** show the microstructure and phase analysis of duplex and ferritic stainless steels after pitting corrosion. As both the stainless steel samples were consolidated by spark plasma sintering method at 1000°C, we can see very less porosity ratios. The black-coloured region in **Figure 11** is pitting-corroded region containing iron oxide. Both the stainless steel samples are having black-coloured region in the microstructure, confirming the pitting corrosion process during electrochemical measurement. All the microstructural analysis was performed to only the stainless steel samples whose corrosion studies were conducted by LSV method at 2 M H_2SO_4 solution.

According to corrosion studies, the rate of corrosion is more in ferritic stainless steel than duplex stainless steel, and it was also confirmed by microstructural analysis.

Phase analysis was carried out to study the volume fraction of iron oxide present in both the stainless steel samples. The presence of iron oxide volume percentage is more in ferritic stainless steel than in duplex stainless steel samples. The volume fraction of iron oxide phase is determined by AxioVision Release software attached to an optical microscope. In **Figure 12**, the red-coloured region corresponds to corroded (iron oxide) stainless steel part, and green colour corresponds to uncorroded stainless steel.

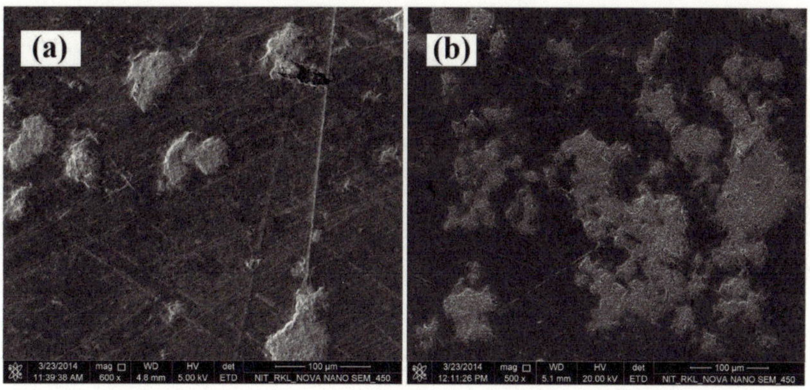

Figure 10.
FESEM images of (a) duplex stainless steel and (b) ferritic stainless steel [30].

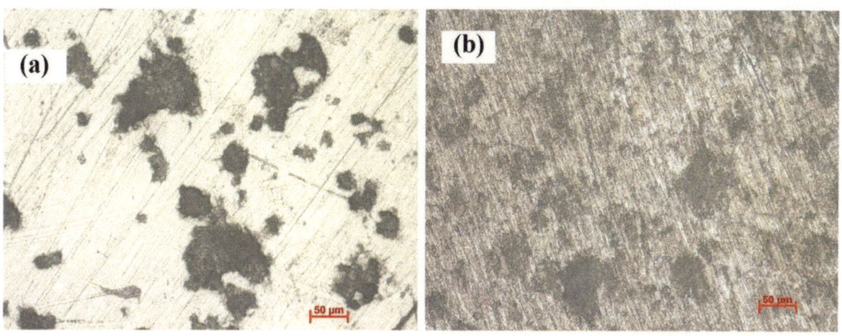

Figure 11.
Optical microscope images of (a) duplex stainless steel and (b) ferritic stainless steel [30].

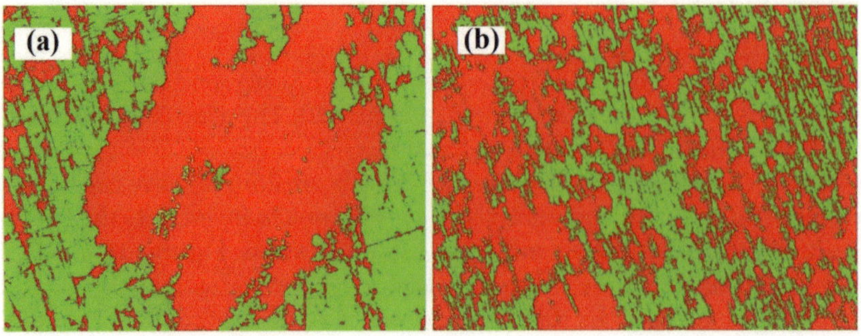

Figure 12.
Optical image phase analysis of (a) duplex stainless steel and (b) ferritic stainless steel using AxioVision Release software (similar to **Figure 11**, *all the images are in a magnification of 50 μm scale bars) [30].*

5. Conclusion of the chapter

In the present chapter, we discussed the pitting corrosion properties of SPS-consolidated stainless steel samples by LSV method at different concentrations of NaCl and H_2SO_4 solutions. As the concentration of both the electrolytes increases from 0.5 to 2 M, then pitting potential of duplex and ferritic stainless steels started to decrease due to the accelerated rate of corrosion reactions at higher concentrations. Microstructural analysis by FESEM and optical microscope shows corroded regions of stainless steel samples. The presence of iron oxide volume percentage is more in ferritic stainless steel than in duplex stainless steel samples. The mechanism of pitting corrosion in H_2SO_4 and NaCl solutions is almost the same; but the only difference is the extent of pitting, pitting potential and pitting current values. Pitting potential value obtained for H_2SO_4 electrolyte is comparatively low compared to the results obtained for NaCl electrolyte. From the results, we can conclude that duplex and ferritic stainless steel samples undergo corrosion easily in the presence of H_2SO_4 than NaCl electrolyte.

Nomenclature

wt.%	weight percentage
MPa	megapascal
°C	degree in celsius
nm	nanometre
μm	micrometre
IP	pitting current
EP	pitting potential
V	voltage
J	current density
M	molar
A	ampere
ip	peak current
A	area of electrode
D	diffusion coefficient
Co	concentration
υ	scan rate

Author details

Shashanka Rajendrachari
Department of Metallurgical and Materials Engineering, Bartin University, Bartin, Turkey

*Address all correspondence to: shashankaic@gmail.com

References

[1] Philip Selvaraj D, Chandramohan P, Mohanraj M. Optimization of surface roughness, cutting force and tool wear of nitrogen alloyed duplex stainless steel in a dry turning process using Taguchi method. Measurement. 2014;**49**:205-215

[2] AK Steel Stainless Steels Corporate Headquarters. AK Steel Corporation 9227 Centre Pointe Drive West Chester, Ohio 45069, (513) 425-5000

[3] Kurzydłowski KJ. Microstructural refinement and properties of metals processed by severe plastic deformation. Bulletin of the Polish Academy of Sciences Technical Sciences. 2004;**52**:275

[4] AustralWright Metals. Stainless Steel—Properties and Applications of Ferritic Grade Stainless Steel. Australia: Austral Wright Metals; 2008

[5] Dobrzanski LA, Brytan Z, Actis Grande M, Rosso M. Properties of duplex stainless steels made by powder metallurgy. Archives of Materials Science and Engineering. 2007;**28**:217-223

[6] Herenu S, Alvarez-Armas I, Armas AF. The influence of dynamic strain ageing on the low cycle fatigue of duplex stainless steel. Scripta Materialia. 2001;**45**:739-745

[7] Miyamoto H, Mirnaki T, Hashimoto S. Super plastic deformation of microspecimens of duplex stainless steel. Materials Science and Engineering A. 2001;**319**:779-783

[8] Schofield MJ, Bradsha R, Cottis RA. Stress corrosion cracking of duplex stainless steel weldments in sour conditions. Materials Performance. 2009;**35**:65-70

[9] Yuan Ma F. Corrosive effects of chlorides on metals. In: Bensalah N, editor. Pitting Corrosion. China: In Tech; 2012. ISBN: 978-953-51-0275-5.

[10] Frankel GS. Pitting corrosion of metals: A review of the critical factors. Journal of the Electrochemical Society. 1998;**145**:2186-2198

[11] Shankar J, Upadhyaya A, Balasubramaniam R. Electrochemical behavior of sintered oxide dispersion strengthened stainless steels. Corrosion Science. 2004;**46**:487-498

[12] Ningshen S, Sakairi M, Suzuki K, Ukai S. The corrosion resistance and passive film compositions of 12% Cr and 15% Cr oxide dispersion strengthened steels in nitric acid media. Corrosion Science. 2014;**78**:322-334

[13] Balaji S, Upadhyaya A. Electrochemical behavior of sintered YAG dispersed 316L stainless steel composites. Materials Chemistry and Physics. 2007;**101**:310-316

[14] Chen J, Matthew Asmussen R, Zagidulin D, Noel JJ, Shoesmith DW. Electrochemical and corrosion behavior of a 304 stainlesssteel-based metal alloy waste form in dilute aqueous environments. Corrosion Science. 2013;**66**:142-152

[15] Li CX, Bell T. Corrosion properties of plasma nitrided AISI 410 martensitic stainless steel in 3.5% NaCl and 1% HCl aqueous solutions. Corrosion Science. 2006;**48**:2036-2049

[16] Metikos-Hukovic M, Babic R, Grubac Z, Petrovic Z, Lajci N. High corrosion resistance of austenitic stainless steel alloyed with nitrogen in an acid solution. Corrosion Science. 2011;**53**:2176-2183

[17] Bard AJ, Faulkner LR. Electrochemical Methods:

Fundamentals and Applications. New York: John Wiley & Sons; 2001

[18] Yan D, Bazant MZ, Biesheuvel PM, Pugh MC, Dawson FP. Theory of linear sweep voltammetry with diffuse charge: Unsupported electrolytes, thin films, and leaky membranes. Physical Review E. 2017;**95**:033303

[19] Compton RG, Laborda E, Ward KR. Understanding Voltammetry: Simulation of Electrode Processes. London: Imperial College Press; 2014

[20] Nahir TM, Clark RA, Bowden EF. Linear-sweep voltammetry of irreversible electron transfer in surface-confined species using the Marcus theory. Analytical Chemistry. 2002;**66**:2595-2598

[21] Randles JEB. Kinetics of rapid electrode reactions. Discussions of the Faraday Society. 1947;**1**:11

[22] Randles JEB. A cathode ray polarograph. Part II.—The current-voltage curves. Transactions of the Faraday Society. 1948;**44**:327

[23] Collect SA. Oscillographic polarography with periodical triangular voltage. Czechoslovak Chemical Communications. 1948;**13**:349-377

[24] Linear Sweep and Cyclic Voltammetry: The Principles. Department of Chemical Engineering and Biotechnology. Available from: https://www.ceb.cam.ac.uk/research/groups/rg-eme/teaching-notes/linear-sweep-and-cyclic-voltametry-the-principles

[25] Shashanka R. Fabrication of nano-structured duplex and ferritic stainless steel by planetary milling followed by consolidation [PhD thesis]. India: NIT Rourkela; 2016

[26] Shashanka R, Chaira D. Phase transformation and microstructure study of nano-structured austenitic and ferritic stainless steel powders prepared by planetary milling. Powder Technology. 2014;**259**:125-136

[27] Shashanka R, Chaira D. Development of nano-structured duplex and ferritic stainless steel by pulverisette planetary milling followed by pressureless sintering. Materials Characterization. 2015;**99**:220-229

[28] Shashanka R, Chaira D. Optimization of milling parameters for the synthesis of nano-structured duplex and ferritic stainless steel powders by high energy planetary milling. Powder Technology. 2015;**278**:35-45

[29] Shashanka R, Chaira D, Kumara Swamy BE. Electrocatalytic response of duplex and yttria dispersed duplex stainless steel modified carbon paste electrode in detecting folic acid using cyclic voltammetry. International Journal of Electrochemical Science. 2015;**10**:5586-5598

[30] Shashanka R, Chaira D, Kumara Swamy BE. Electrochemical investigation of duplex stainless steel at carbon paste electrode and its application to the detection of dopamine, ascorbic and uric acid. International Journal of Scientific & Engineering Research. 2015;**6**:1863-1871

[31] Gupta S, Shashanka R, Chaira D. Synthesis of nano-structured duplex and ferritic stainless steel powders by planetary milling: An experimental and simulation study. 4th National Conference on Processing and Characterization of Materials. IOP Conference Series: Materials Science and Engineering. 2015;**75**:012033

[32] Shashanka R, Chaira D. Effects of nano-Y_2O_3 and sintering parameters on the fabrication of PM duplex and ferritic stainless steels. Acta Metallurgica Sinica (English Letters). 2016;**29**:58-71

[33] Nayak AK, Shashanka R, Chaira D. Effect of nanosize yttria and tungsten addition to duplex stainless steel during high energy planetary milling. 5th National Conference on Processing and Characterization of Materials. IOP Conference Series: Materials Science and Engineering. 2016;**115**. DOI: 012008

[34] Shashanka R, Chaira D, Kumara Swamy BE. Fabrication of yttria dispersed duplex stainless steel electrode to determine dopamine, ascorbic and uric acid electrochemically by using cyclic voltammetry. International Journal of Scientific & Engineering Research. 2016;7:1275-1285

[35] Shashanka R, Chaira D. Effect of sintering temperature and atmosphere on non-lubricated sliding wear of nano-yttria dispersed and yttria free duplex and ferritic stainless steel fabricated by powder metallurgy. Tribology Transactions. 2017;**60**:324-336

[36] Shashanka R, Chaira D, Chakravarty D. Fabrication of nano-yttria dispersed duplex and ferritic stainless steels by planetary milling followed by spark plasma sintering and non-lubricated sliding wear behaviour study. Journal of Materials Science and Engineering B. 2016;**6**:111-125

[37] Shashanka R. Synthesis of nano-structured stainless steel powder by mechanical alloying—An overview. International Journal of Scientific & Engineering Research. 2017;**8**:588-594

[38] Shashanka R, Chaira D, Kumara Swamy BE. Effect of Y_2O_3 nanoparticles on corrosion study of spark plasma sintered duplex and ferritic stainless steel samples by linear sweep voltammetric method. Archives of Metallurgy and Materials. 2018;**63**:745-759

[39] Shashanka R. Effect of sintering temperature on the pitting corrosion of ball milled duplex stainless steel by using linear sweep voltammetry. Analytical & Bioanalytical Electrochemistry. 2018;**10**:349-361

[40] Sulphuric Acid on the Web. Knowledge for the Sulphuric Acid Industry. DKL Engineering, Inc. Corrosion. June 6, 2005

[41] Tamarit EB, García-Garcia DM, Garcia Anton J. Imposed potential measurements to evaluate the pitting corrosion resistance and the galvanic behaviour of a highly alloyed austenitic stainless steel and its weldment in a LiBr solution at temperatures upto 150°C. Corrosion Science. 2011;**53**:784

Chapter 6

Carbon-Based Electrode Application for Determination and Differentiation of Chromium Ion Species Using Voltammetric Method

Santhy Wyantuti, Safri Ishmayana, Diana Rakhmawaty Eddy and Yeni Wahyuni Hartati

Abstract

Cr(III) and Cr(VI) have significantly different toxicological properties, with the latter being toxic for human health. The anodic stripping voltammetry using pencil graphite electrode (PGE) was developed to determine Cr(III), and it can measure the ion within linear range 12.5–75 ppm with 0.31 ppm detection limit, while adsorptive striping voltammetry was used to determine very low level of Cr(VI) using PGE with the addition of diethylenetriaminepentaacetic acid (DTPA). The results indicate that Cr(VI) can give electrochemical response when adsorptive stripping voltammetry was used, and addition of DTPA gives a specific Cr(VI) peak at −0.65 V. A more advanced method was developed for the differentiation of Cr(VI) and Cr(III) using gold nanoparticle-modified glassy carbon electrode (AuNP-GCE). The AuNP-GCE was applied as electrochemical sensor for Cr(VI) and Cr(III) analysis. The electrode can differentiate Cr(VI) from Cr(III) using different conditions. Linear range for Cr(VI) determination using the method was between 0.050 and 0.250 μg/L, with detection limit of 2.38 ng/L, while for Cr(III), the linear range was between 0.500 and 75.000 ng/L, with detection limit of 0.010 ng/L. The study of other ions' effects on the method showed that copper(II), cadmium(II), nickel(II), and iron(III) do not interfere the measurement.

Keywords: Cr(III), Cr(IV), glassy carbon electrode, gold nanoparticle, pencil graphite electrode, voltammetry

1. Introduction

Chromium occurs in nature in two valencies, Cr(III) and Cr(VI). The two ionic species have distinct toxicity and reactivity [1]. Cr(VI) has high toxicity, while Cr(III) is less toxic, and the latter is also known to be one of the essential ions for humans, which is required in particular amount in the diet [2]. Due to their opposite nature in toxicity, it is important to distinguish between Cr(III) and Cr(VI). Therefore, specific and sensitive determination methods to distinguish both of the species are required.

Since Cr(III) belongs to essential nutrition ion required for maintaining normal physiological function, thus trace amount determination of Cr(VI) becomes very important to avoid toxic intake in the diets [3]. There have been several sensitive methods used for Cr(III) and Cr(VI) determination including atomic absorption spectrophotometry (AAS) [4, 5], plasma mass spectroscopy [6], spectrofluorimetry [7], chemiluminescence [8], spectrophotometry [9], and electrochemical methods [10–12]. Determination of chromium ions using voltammetric method is evolving along with the requirements to accurately measure the ionic species [13–18].

The material of working electrode influences the performance of voltammetric method. Application of carbon-based electrode utilization is currently growing in electroanalytical field because it has some advantages including wide electrical potential range, low background current, inexpensive, inert, and suitable for a wide variety of sensors [19]. Modification of the electrode is being explored to enhance the electrode performance [20, 21]. The modified electrode needs to be evaluated for their performance. In the present chapter, investigation of carbon-based electrode without and with modification was evaluated for their performance to determine Cr(VI) and Cr(III).

2. Anodic stripping voltammetry with pencil graphite electrode for determination of Cr(III)

For determination of Cr(III) using graphite pencil electrode (GPE), three solvents were trialed [21], i.e., 0.01 M hydrochloric acid, acetate buffer pH 5.0, and double distilled water. Potential range from -800 to 1000 mV with 90 s deposition time, 10 s cleaning time, and 20 mV/s scan rate was used for the measurement using anodic stripping voltammetry. This experiment was performed to measure the background signal in order to get a correction factor which can avoid the presence of undesired peak current from the solvent. No peak was detected when 0.01 M hydrochloric acid was used at the condition applied, while acetate buffer pH 5.0 and double distilled water gave relatively high peak current which can interfere the measurement (data not shown). Therefore, hydrochloric acid was chosen as solvent for the method. For further experiment, Cr(III) solution was made by dissolving Cr(III) chloride in 0.01 M hydrochloric acid as solvent.

An experiment for investigating the linearity of the method was performed by measuring Cr(III) within 1–500 ppm range. The experiment was performed using a potential range of −1.5 to +1.0 V with 60 s of deposition time, 10 s of cleaning time, and 50 mV/s of scan rate. The electrode used in the experiment was pencil graphite electrode, and the linear concentration range was determined by correlating height of peak current and concentration of Cr(III). The linear concentration range was found in 12.5–75 ppm range, and the result is presented in **Figure 1**.

The range of linear concentration at 12.5–75 ppm gave correlation coefficient of 0.9936, which indicate good relationship with correlation equation of $y = 0.1778x + 0.747$. PGE can be used as working electrode for determination of Cr(III) with optimum condition achieved at 60 s of deposition time and 50 mV/s of scan rate. Linear range of Cr(III) measurement with the proposed condition was 12.5–75 ppm with 0.31 ppm limit of detection.

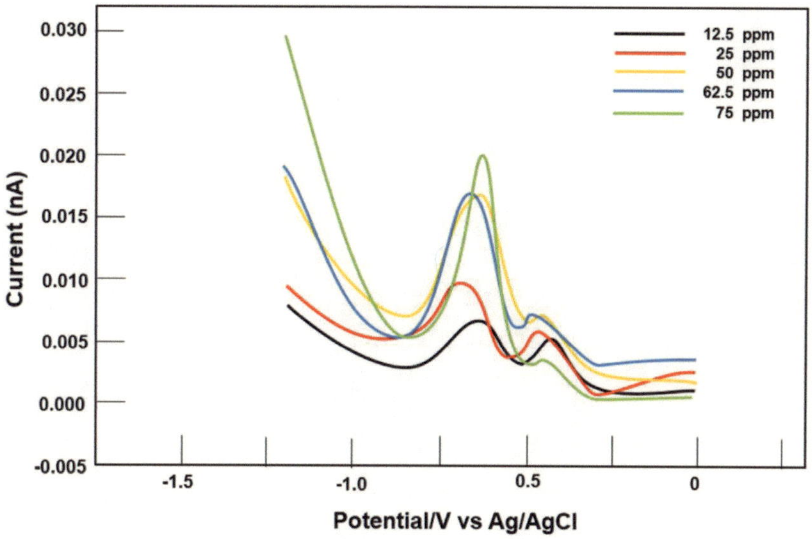

Figure 1.
Typical peak response of chromium(III) voltammogram using PGE. Experiment condition: solvent = 0.01 M hydrochloric acid, chromium(III) concentration = 12.5, 25, 50, 62.5, and 75 ppm was performed sequentially with 60 s of deposition time, 50 mV/s of scan rate, and deposition potential at −1260 mV [21].

3. The development of graphite pencil electrode for Cr(VI) determination using adsorptive stripping voltammetry

The addition of a chelating agent in the form of ligands to the Cr(VI) signal measurements has been done in several previous studies such as addition of rubeanic acid ligand [22] and addition of diethylenetriaminepentaacetic acid (DTPA) as ligand with electrolyte support of morpholinoethanesulfonate acid buffer (MES) [23]. From previous studies it can be concluded that the addition of a ligand for the measurement of Cr(VI) signal by voltammetry can increase the measurement sensitivity [24]. **Figure 2** shows that after the addition of the DTPA, the peak generated on the voltammogram becomes specific, and no further undesirable impurity peaks occur (black line designated as d). After background correction, it can be concluded that with the addition of DTPA, the impurity peaks were reduced, a specific peak on the voltammogram was produced, and the peak current height of chromium increases so that the measurement sensitivity also increases.

Calibration curve was prepared by measuring differential pulse adsorptive stripping voltammetry of Cr(VI) with the following concentrations: 1, 10, 20, and 30 ppm. The voltammetric measurement was carried out at a potential deposition condition of +810 mV, a scan rate of 50 mV/s, and a deposition time of 60 s with stirring using a magnetic stirrer.

From **Figure 3** it can be seen that the peak current height increases with increasing analyte concentration in the voltammetric cell. This may be due to the increasing number of Cr species deposited on the working electrode when the analytical concentration is added, thereby increasing the peak current in the voltammogram. The linear regression of the standard curve is $y = 0.6319x + 8.5796$ with correlation coefficient of 0.957.

The measurement of Cr(VI) peak current was performed triplicate to calculate precision value. For 1 ppm, a fairly low standard deviation was obtained which is 0.3466. This indicates that the measurement results have a fairly good accuracy or precision, and from the calculation, the coefficient of variation of

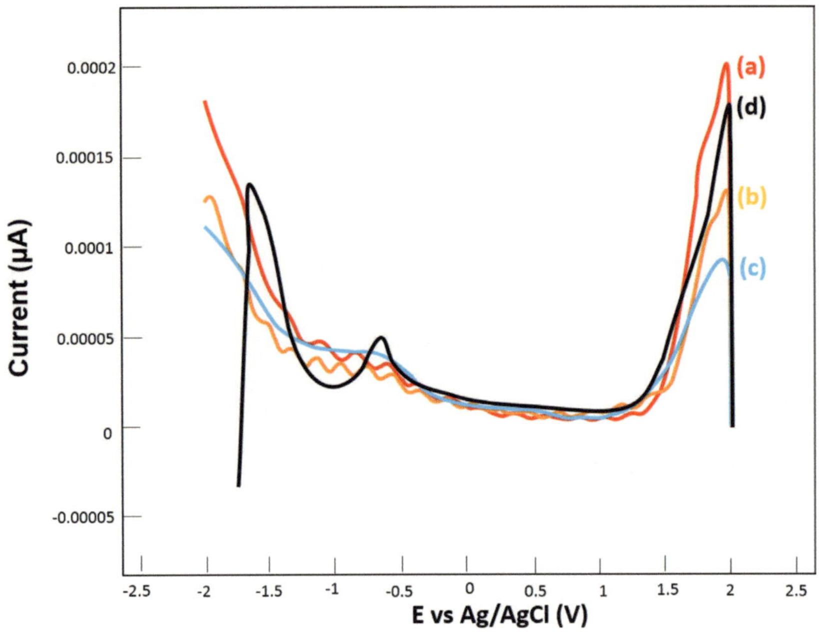

Figure 2.
Voltammogram of (a) 0.2 M acetate buffer pH 6.0, (b) DTPA 868 ppm, (c) 30 ppm Cr(VI), and (d) 30 ppm Cr(VI) with addition of DTPA. Measurement conditions: potential deposition, +810 mV; deposition time, 60 s; scan speed, 50 mV/s; potential range, +2000 V to −2000 mV.

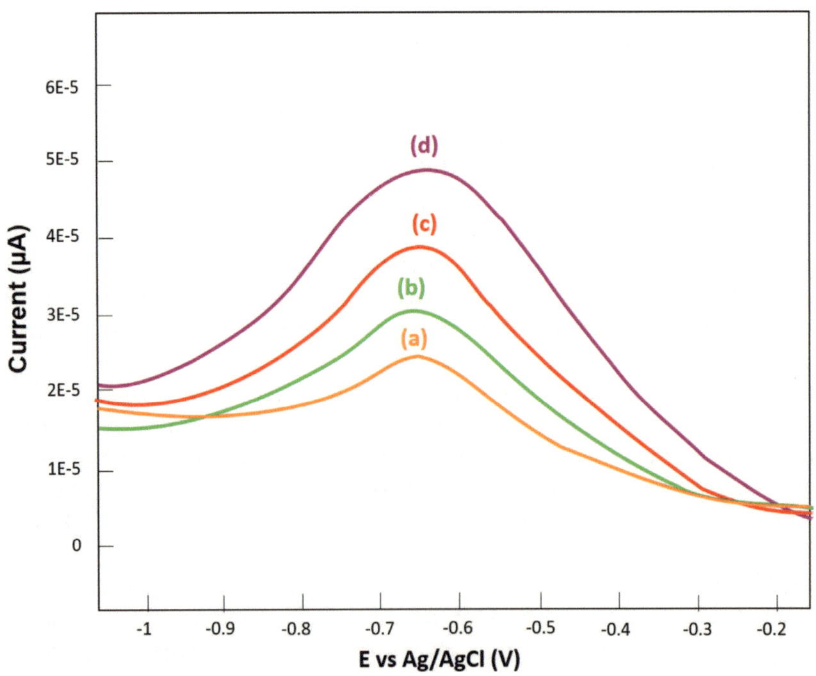

Figure 3.
Voltammogram of Cr(VI) using PGE with concentration of (a) 1, (b) 10, (c) 20, and (d) 30 ppm with DTPA addition. Measurement conditions: potential deposition, +810 mV; deposition time, 60 s; scan rate, 50 mV/s; potential range, +2000 V to −2000 mV.

4.56% was also obtained. The addition of DTPA as ligand can be used to detect Cr(VI) and can produce a specific peak for Cr(VI) at a potential region of −0.6474 V.

4. Voltammetric determination of Cr(VI) using gold nanoparticle-modified glassy carbon electrode

AuNP was synthesized using citric ion as capping agent. After reduction using $NaBH_4$ solution, the gold solution color was changed from yellow to red ruby (λ_{max} = 518 nm), which indicate that AuNP colloid was formed with average dimension of 2–5 nm [25]. Characterization using TEM indicates that the average diameter of the colloid was ~2.2 nm (**Figure 4**). This gold nanoparticle was then used to modify the GCE.

GCE was modified by attaching AuNP onto the electrode surface. Modification of GCE was initiated by polishing the electrode followed by activation. This step was performed so that the electrode can be covered evenly and maximally by the AuNP, which eventually can increase sensitivity when used to determine the analyte. Photochemical reaction is one of the many methods to activate the GCE surface. The photochemical reaction was performed by immersing the electrode into concentrated ammonium hydroxide and followed by irradiation under UV light for 6 hours at 254 nm [26]. This process is known as self-assembly which can substitute hydrogen group (–H) with amine group ($-NH_2$) of the electrode. Due to the nature of the amine group, more AuNP can be attached to the electrode and improve binding to the electrode [27], as indicated by SEM-EDS data. SEM-EDS data demonstrated that the self-assembly process was better than the adsorption process. Modification of GCE by self-assembly process reached 11.55% of AuNP covering the electrode (**Figure 5**), while the adsorption process (direct immersion of the electrode) only reached 2.05% covering according to SEM-EDS data.

From the present study, it was found that optimum deposition potential and deposition time for determination of 0.5 μg/L Cr(VI) in 0.01 M hydrochloric acid as

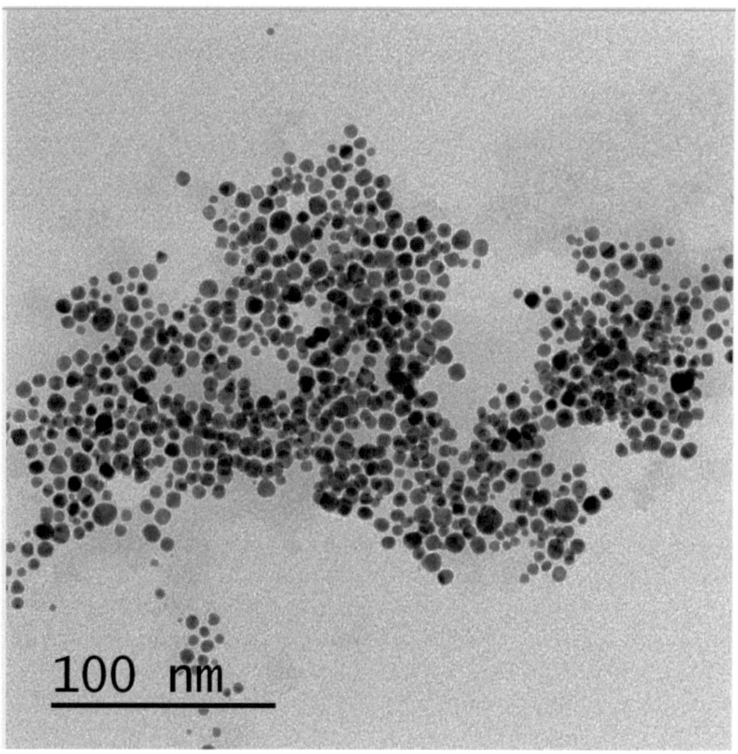

Figure 4.
TEM image of gold nanoparticles.

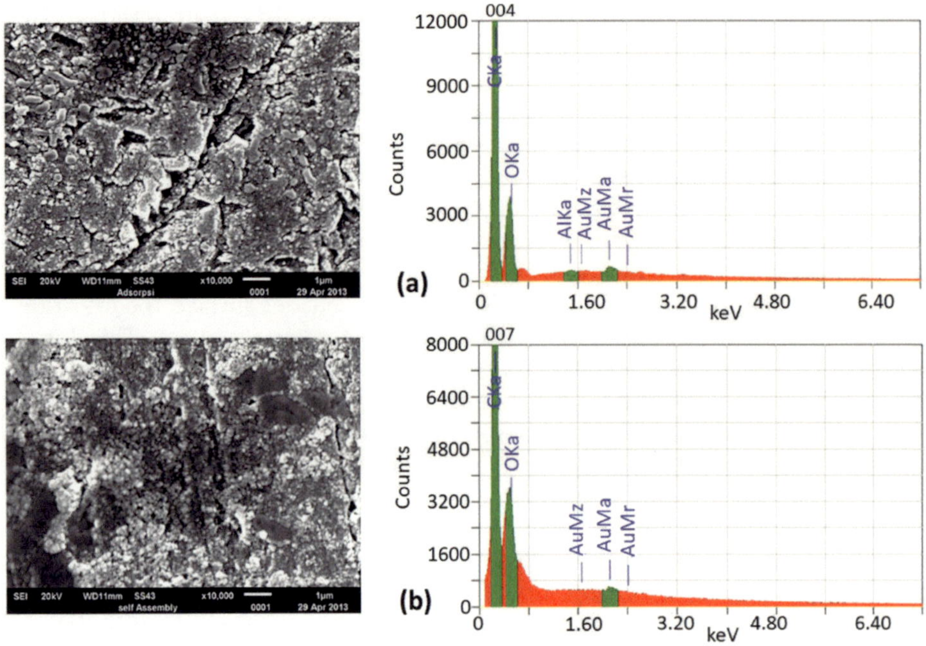

Figure 5.
SEM image of gold nanoparticle-modified glassy carbon electrode surface. Modification was conducted using (a) adsorption and (b) self-assembly process.

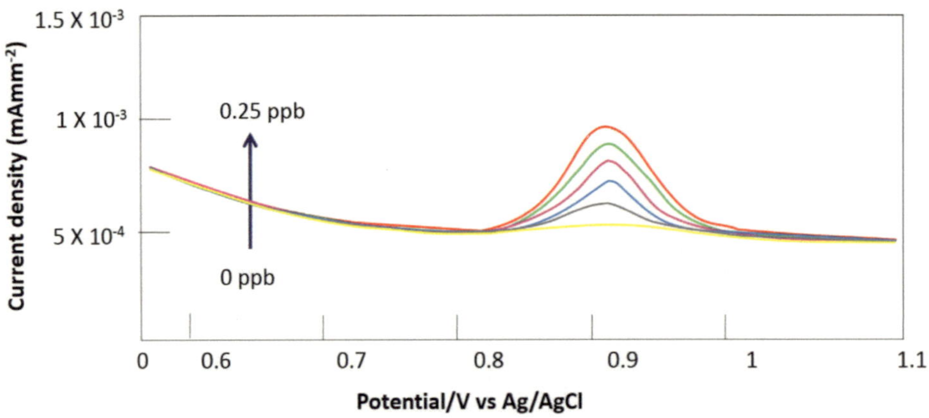

Figure 6.
Voltammogram of various chromium(VI) concentrations (0–0.25 µg/L) with AuNP-GCE. The condition of measurement is as follows: deposition potential, deposition time, scan rate, and amplitude modulation at 800 mV, 120 s, 50 mV/s, and 500 mV, respectively.

supporting electrolyte solution were 800 mV and 120 s, respectively. The optimum condition was then applied to develop a calibration curve with concentration ranging from 0 to 25 µg/L, and the current density to potential data of the concentrations is presented in **Figure 6**. From the data, a linear calibration curve (r^2 = 0.9896) was developed and is presented on **Figure 7** with correlation coefficient and linear regression equation of 0.9911 and $I = 2.25 \times 10^{-3} C + 5.89 \times 10^{-4}$, respectively. Through calculation, it was obtained that limit of detection and precision of the method are 2.38 ng/L and 99.14%, respectively.

To see whether we can distinguish Cr(VI) from Cr(III) using condition described earlier, we measure individual and mixture of the ionic species. The result

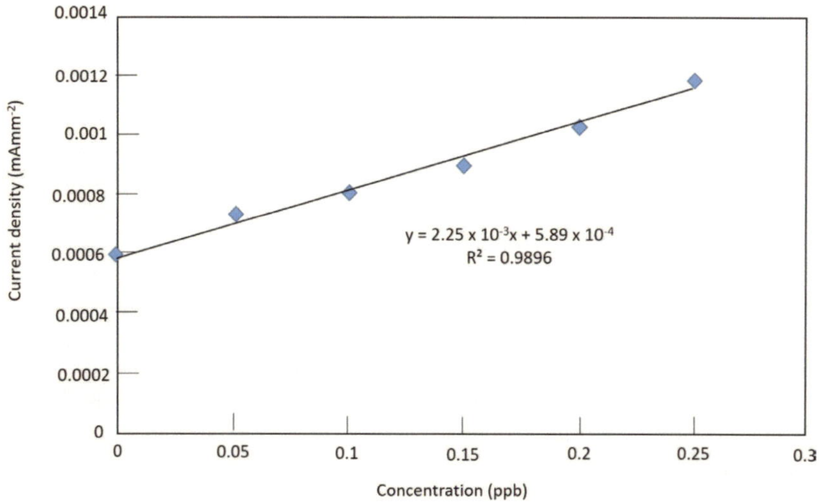

Figure 7.
Calibration curve of various chromium(VI) concentrations with deposition potential, deposition time, scan rate, and amplitude modulation at 800 mV, 120 s, 50 mV/s, and 500 mV, respectively.

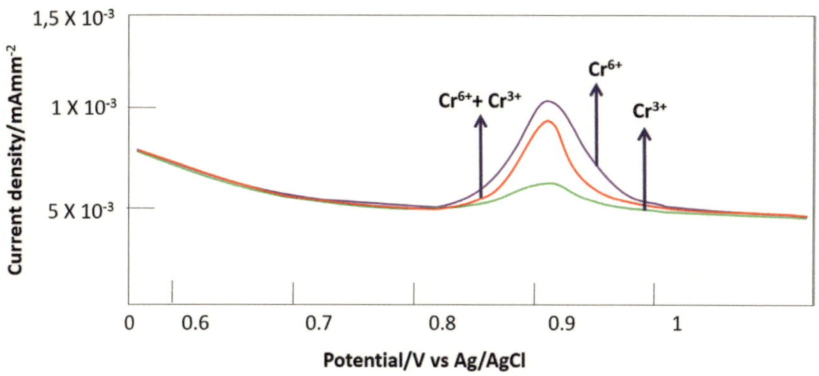

Figure 8.
Voltammogram of Cr(III) and Cr(VI) and mixture of Cr(III) and Cr(VI) in 0.01 M hydrochloric acid using AuNP-GCE against Ag/AgCl with deposition potential of 800 mV, deposition time of 120 s, and amplitude modulation of 0.05 V.

is presented in **Figure 8**. In contrast with Cr(VI), it was found that in this condition Cr(III) does not give a sharp peak. However, it seems that the peak of the mixture was slightly reduced compared to Cr(VI) peak. Therefore, we can measure Cr(VI) in this condition with a neglected effect from Cr(III).

5. Voltammetric detection of Cr(III) using gold nanoparticle-modified glassy carbon electrode

To determine the optimum condition for determination of Cr(III) using AuNP-GCE, a preliminary experiment was conducted by varying deposition potential. The potential used were ranging from −2.0 to −1.0 V with scan rate and amplitude modulation of 0.05 V/s and 0.05 V, respectively. The highest efficiency for Cr(III) 500 ppb in 0.2 M acetate buffer was found at −1.5 V, while optimization of deposition time leads us to conclude that the best result was obtained at 90 s [17]. The optimized parameters are very crucial in stripping step that have pronounced effect on both sensitivity and linear range. Therefore, from the preliminary experiment, we conclude that the optimum condition for determination of Cr(III) using AuNP-GCE is using −1.5 V of deposition potential and 90 s of deposition time.

The optimum condition was then applied to prepare a calibration curve of Cr(III) using AuNP-GCE differential pulse stripping voltammetry. The calibration curve was obtained by varying concentration of Cr(III) and measuring it at −1.5 V of deposition potential, 90 s of deposition time, 0.05 V/s of scan rate, and 0.05 V of amplitude modulation. The voltammetric measurement and calibration curve are presented in **Figure 9**. By calculation, a linear regression equation and correlation coefficient of y = 0.142 + 0.004x and r = 0.9955, respectively, were obtained. The result showed that the linear range was obtained at 0.5–75 ppb range. The reproducibility of the method was then determined from the replicate measurement of 1 ppb Cr(III). The calculation results in a mean of current response of 0.14 μA ppb^{-1}, which is referred to as sensitivity of the method. It corresponds to the value of relative standard deviation (RSD) of 2.4%. Based on calculation suggested by Miller and Miller [28], limit of detection of the method is 0.01 ppb, which is three times of the value of standard deviation baseline. Utilization of AuNP-modified electrode improves at least one order of magnitude detection limit compared to voltammetric method using film electrode. Limited publication of voltammetric detection of Cr(III) compared to Cr(VI) and poor sensitivity of non-gold electrode lead us to infer that gold or gold-modified electrodes are required to get high-quality Cr(III) voltammetric profiles. It is previously reported that high-quality voltammetric profiles for Cr(III) appear when gold electrode is used [29].

Optimum condition was then applied for determination of Cr(III) in electroplating wastewater, which is a more complex matrix than the standard solution. Based on standard addition method, we found that the concentration of Cr(III) in the sample was 0.146 ppm, corresponding to 97% recovery and in good agreement with the result from AAS method (AAnalyst 400, PerkinElmer) after appropriate

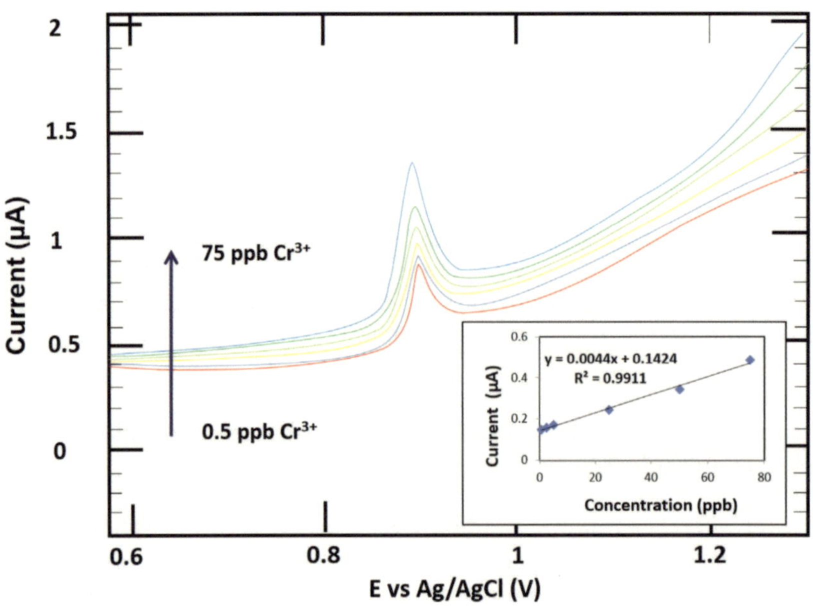

Figure 9.
Voltammetric performance for Cr(III) concentration variation. Inset shows corresponding calibration curve at an optimum condition.

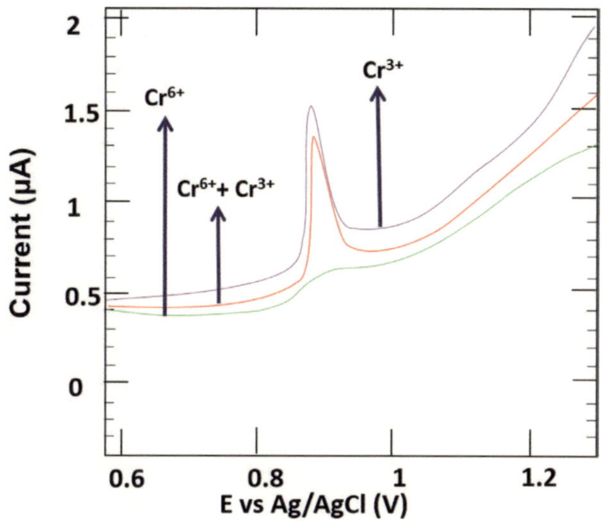

Figure 10.
Voltammogram Cr(III) and Cr(VI) and mixture of Cr(III) and Cr(VI) in acetate buffer pH 5 using AuNP-GCE electrode against Ag/AgCl electrode with deposition potential of −1500 mV, deposition time of 90 s, and amplitude modulation of 50 mV.

extraction. Therefore, the proposed voltammetric method using the fabricated electrochemical AuNP-GCE sensor gives satisfactory result and is applicable for determination of Cr(III) in complex matrix, such as wastewater.

There is a possibility that Cr(VI) interferes the Cr(III) measurement in the proposed condition. Therefore, we measure the voltammogram of individual and mixture of the ionic species. The results are presented in **Figure 10**. From the figure, it is clear that by using the proposed condition, clear peak was only detected for Cr(III) species, while Cr(VI) did not show any significant peak at the corresponding potential. Mixture of Cr(III) and Cr(VI) also gives similar peak which corresponds to the individual Cr(VI) peak.

6. Study on interfering ions

In order to evaluate the effect of foreign metal ions dissolved in the media, solutions containing 50 ppb of Cr(III) and each of the following metals Cd(II), Cu(II), Zn(II), Cr(VI), Ni(II), and Fe(III) all at the same concentration of 50 ppb were prepared. Measurement was conducted at an optimum condition described in Subchapter 4, and the result is presented in **Table 1**. The effect of interfering metal ion on Cr(III) was indicated by the change in peak current signal during measurement. The addition of foreign ions does not influence the peak current corresponding to Cr(III). The percentage variations of the peak current induced by the presence of interfering ions with respect to Cr(III) alone are 1.02% for Cd(II), 0.88% for Cu(II), 0.29% for Zn(II), 3.70% for Cr(VI), 0.07% for Ni(II), and 0.40% for Fe(III). These deviations from Cr(III) measurement at 0.9470 V and 1.360 μA remain acceptable, and, therefore, the accurate detection of Cr(III) is still possible under these conditions.

Cd(II) and Zn(II) ions have a standard reduction potential similar to Cr(III) ion. During the current measurement of Cr(III) with the presence of interfering ions, they also deposited on the electrode surface, reducing the amount of Cr(III) on the electrode surface. At the deposition stage, when −1.5 V electric potential was

Interfering metal ion	Potential (V)	Current (μA)	Change (%)
None	0.9470	1.360	–
Cd(II)	0.8857	1.374	1.02
Cu(II)	0.8357	1.348	0.88
Zn(II)	0.8507	1.356	0.29
Cr(VI)	0.9359	1.323	3.7
Ni(II)	0.9100	1.361	0.07
Fe(III)	0.8507	1.356	0.4

Table 1.
Result of measurement of 50 ng/L Cr(III) in the presence of 50 ng/L interfering metal ions.

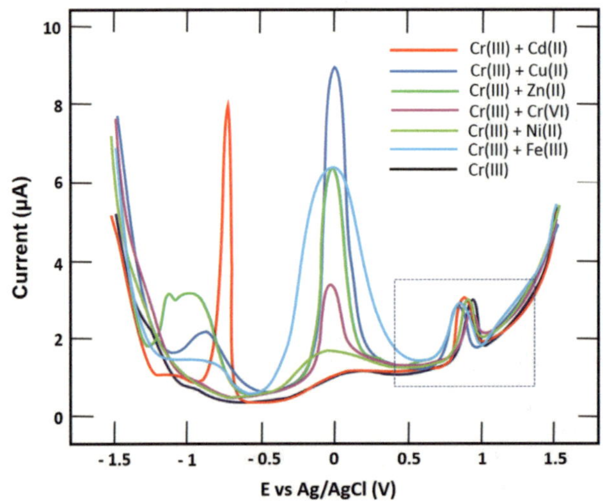

Figure 11.
Differential pulse voltammogram of Cr(III) using AuNP-GCE with −1.5 V of deposition potential, 90 s of deposition time, 0.05 V/s of scan rate, and 0.05 V of amplitude modulation against Ag/AgCl. Concentration of Cr(III) was 50 ppb in acetate buffer pH 5 without and with the presence of Cd(II), Cu(II), Zn(II), Cr(VI), Ni(II), or Fe(III) as interfering ions with concentration of 50 ppb.

applied, the ions will be reduced too because their standard reduction potential are more positive than the deposition potential applied. The effect of interfering metal ion is presented in **Figure 11**.

Fe(III), Cr(VI), Cu(II), and Ni(II) ions have more positive standard reduction potential than Cr(III), and therefore when −1.5 V potential was applied, the interfering ions were also reduced and causing less Cr(III) reduced than it should be. Therefore, the peak of Cr(III) is reduced, while Zn(II) and Cd(II) have more negative standard reduction potential than Cr(III), and therefore when reduction potential was applied, Cr(III) will be reduced followed, in order, by Zn(II) and Cd(II).

7. Conclusions

Carbon-based electrode can be applied as working electrode for the determination of Cr(III) and Cr(VI). PGE can detect the presence of Cr(III) in solution, while for Cr(VI) addition of DTPA is needed to get a specific peak current of Cr(VI). The two methods utilizing PGE can detect Cr(III) and Cr(VI) in ppm scale. To enhance the analytical performance, AuNP-GCE was used as working electrode, and it showed better results. The modified electrode can measure both Cr(III) and Cr(VI) in ppb scale. By using a different condition, we succeeded to measure Cr(III) and Cr(VI) with the same working electrode.

Acknowledgements

The authors would like to acknowledge the financial support from the Academic Leadership Grant Program, Universitas Padjadjaran, and the Directorate of Research and Community Service through Superior Research of Higher Education (DRPM-PDUPT).

Author details

Santhy Wyantuti*, Safri Ishmayana, Diana Rakhmawaty Eddy and Yeni Wahyuni Hartati
Department of Chemistry, Faculty of Mathematics and Natural Sciences, Universitas Padjadjaran, Sumedang, West Java, Indonesia

*Address all correspondence to: santhy.wyantuti@unpad.ac.id

References

[1] Cespón-Romero RM, Yebra-Biurrun MC, Bermejo-Barrera MP. Preconcentration and speciation of chromium by the determination of total chromium and chromium(III) in natural waters by flame atomic absorption spectrometry with a chelating ion-exchange flow injection system. Analytica Chimica Acta. 1996;**327**(1):37-45. DOI: 10.1016/0003-2670(96)00062-1

[2] Costa M, Klein CB. Toxicity and carcinogenicity of chromium compounds in humans. Critical Reviews in Toxicology. 2006;**36**(2):155-163. DOI: 10.1080/10408440500534032

[3] Domínguez-Renedo O, Alonso-Lomillo MA, Arcos-Martinez MJ. Optimisation procedure for the inhibitive determination of chromium(III) using an amperometric tyrosinase biosensor. Analytica Chimica Acta. 2004;**521**(2):215-221. DOI: 10.1016/j.aca.2004.06.026

[4] Bağ H, Türker AR, Lale M, Tunçeli A. Separation and speciation of Cr(III) and Cr(VI) with *Saccharomyces cerevisiae* immobilized on sepiolite and determination of both species in water by FAAS. Talanta. 2000;**51**(5):895-902. DOI: 10.1016/S0039-9140(99)00354-9

[5] Kiran K, Kumar KS, Prasad B, Suvardhan K, Lekkala RB, Janardhanam K. Speciation determination of chromium(III) and (VI) using preconcentration cloud point extraction with flame atomic absorption spectrometry (FAAS). Journal of Hazardous Materials. 2008;**150**(3):582-586. DOI: 10.1016/j.jhazmat.2007.05.007

[6] Hirata S, Honda K, Shikino O, Maekawa N, Aihara M. Determination of chromium(III) and total chromium in seawater by on-line column preconcentration inductively coupled plasma mass spectrometry. Spectrochimica Acta Part B: Atomic Spectroscopy. 2000;**55**(7):1089-1099. DOI: 10.1016/S0584-8547(00)00169-5

[7] Tang B, Yue T, Wu J, Dong Y, Ding Y, Wang H. Rapid and sensitive spectrofluorimetric determination of trace amount of Cr(III) with o-vanillin-8-aminoquinoline. Talanta. 2004;**64**(4):955-960. DOI: 10.1016/j.talanta.2004.04.016

[8] Gammelgaard B, Y-p L, Jøns O. Improvement on simultaneous determination of chromium species in aqueous solution by ion chromatography and chemiluminescence detection. Analytica Chimica Acta. 1997;**354**(1-3):107-113. DOI: 10.1016/S0003-2670(97)00421-2

[9] Kaneko M, Kurihara M, Nakano S, Kawashima T. Flow-injection determination of chromium(III) by its catalysis on the oxidative coupling of 3-methyl-2-benzothiazolinone hydrazone with *N*-ethyl-*N*-(2-hydroxy-3-sulfopropyl)-3-methoxyaniline. Analytica Chimica Acta. 2002;**474**(1-2):167-176. DOI: 10.1016/S0003-2670(02)01005-X

[10] Chatzitheodorou E, Economou A, Voulgaropoulos A. Trace determination of chromium by square-wave adsorptive stripping voltammetry on bismuth film electrodes. Electroanalysis. 2004;**16**(21):1745-1754. DOI: 10.1002/elan.200403051

[11] Husáková L, Bobrowski A, Šrámková J, Królicka A, Vytřas K. Catalytic adsorptive stripping voltammetry versus electrothermal atomic absorption spectrometry in the determination of trace cobalt and chromium in human urine. Talanta. 2005;**66**(4):999-1004. DOI: 10.1016/j.talanta.2005.01.003

[12] Safavi A, Maleki N, Shahbaazi HR. Indirect determination of hexavalent chromium ion in complex matrices by adsorptive stripping voltammetry at a mercury electrode. Talanta. 2006;**68**(4):1113-1119. DOI: 10.1016/j.talanta.2005.07.015

[13] Campbell FW, Compton RG. The use of nanoparticles in electroanalysis: An updated review. Analytical and Bioanalytical Chemistry. 2010;**396**(1):241-259. DOI: 10.1007/s00216-009-3063-7

[14] Hong J, Wang W, Huang K, Yang W-Y, Zhao Y-X, Xiao B-L, et al. A highly efficient nano-cluster artificial peroxidase and its direct electrochemistry on a nano complex modified glassy carbon electrode. Analytical Sciences. 2012;**28**(7): 711-716. DOI: 10.2116/analsci.28.711

[15] Jin W, Wu G, Chen A. Sensitive and selective electrochemical detection of chromium(VI) based on gold nanoparticle-decorated titania nanotube arrays. Analyst. 2014;**139**(1):235-241. DOI: 10.1039/C3AN01614E

[16] Saha K, Agasti SS, Kim C, Li X, Rotello VM. Gold nanoparticles in chemical and biological sensing. Chemical Reviews. 2012;**112**(5): 2739-2779. DOI: 10.1021/cr2001178

[17] Wyantuti S, Hartati YW, Firdaus ML, Panatarani C, Tjokronegoro R. Fabrication of gold nanoparticles-modified glassy carbon electrode and its application for voltammetric detection of Cr(III). International Journal of Scientific & Technology Research. 2015;**4**(1):135-139

[18] Wyantuti S, Tjokronegoro R, Hartati YW, Panatarani C. Cyclic voltammetry study of chromium(VI) and chromium(III) at a gold nanoparticles-modified glassy carbon electrode. Procedia Chemistry. 2015;**17**:170-176. DOI: 10.1016/j.proche.2015.12.109

[19] Wang J. Analytical Electrochemistry. 3rd ed. New Jersey: John Wiley & Sons; 2006

[20] Bui M-PN, Li CA, Han KN, Pham X-H, Seong GH. Simultaneous detection of ultratrace lead and copper with gold nanoparticles patterned on carbon nanotube thin film. Analyst. 2012;**137**(8):1888-1894. DOI: 10.1039/C2AN16020J

[21] Wyantuti S, Hafidza RA, Ishmayana S, Hartati YW. Anodic stripping voltammetry with pencil graphite electrode for determination of chromium(III). Journal of Physics: Conference Series. 2017;**812**:012006. DOI: 10.1088/1742-6596/812/1/012006

[22] Abbasi S, Bahiraei A. Ultra trace quantification of chromium(VI) in food and water samples by highly sensitive catalytic adsorptive stripping voltammetry with rubeanic acid. Food Chemistry. 2012;**133**(3):1075-1080

[23] Korolczuk M, Stępniowska A. Determination of Cr(VI) in the presence of high excess of a Cr(III) by adsorptive stripping voltammetry. Talanta. 2012;**88**:427-431. DOI: 10.1016/j.talanta.2011.11.012

[24] Wyantuti S, Anjani R, Hartati YW. The development of graphite pencil electrode for chromium(VI) determination using adsorptive stripping voltammetry. In: Poceeding of the Bengkulu International Conference on Science and Education. Bengkulu: Universitas Bengkulu; 14-15 December 2017. p. 310-313

[25] Frens G. Controlled nucleation for the regulation of the particle size in monodisperse gold suspensions. Nature Physical Science. 1973;**241**(105):20-22

[26] Wyantuti S, Ishmayana S, Hartati YW. Voltammetric determination of Cr(VI) using gold nanoparticles-modified glassy carbon electrode. Procedia Chemistry. 2015;**16**:15-23. DOI: 10.1016/j.proche.2015.12.010

[27] Tian R-H, Rao TN, Einaga Y, Zhi J-F. Construction of two-dimensional arrays gold nanoparticles monolayer onto boron-doped diamond electrode surfaces. Chemistry of Materials. 2006;**18**(4):939-945. DOI: 10.1021/cm0519481

[28] Miller J, Miller JC. Statistics and Chemometrics for Analytical Chemistry. 6th ed. Canada: Pearson Education; 2010. p. 296

[29] Welch CM, Hyde ME, Nekrassova O, Compton RG. The oxidation of trivalent chromium at polycrystalline gold electrodes. Physical Chemistry Chemical Physics. 2004;**6**(12):3153-3159. DOI: 10.1039/B404243C